# EINSTEIN'S FALLACY

# EINSTEIN'S FALLACY
## THE END OF TIME DILATION

How a single word misinterpreted by Einstein
has fooled all scientists in the world
and led to a major fallacy unique in scientific history

# JOHN DOAN

Written, Designed, and Illustrated by John Doan
Cover: 'Blue Einstein' – oil on canvas – Picasso cover

.

 A catalogue record for this
NATIONAL    book is available from the
LIBRARY    National Library of Australia
OF AUSTRALIA

ISBN: 978-0-9750421-1-3 (paperback)
ISBN: 978-0-9750421-2-0 (hardcover)
ISBN: 978-0-9750421-3-7 (ebook)

To a schoolboy who never stops searching.

# CONTENTS

# LIST OF ILLUSTRATIONS

# PROLOGUE
## WHEN A CHILD IS BORN

*Think different.*
*STEVE JOBS*

I often work with children with severe intellectual disability. Down syndrome, cerebral palsy, fragile X, autism. Most have delayed speech, difficulty thinking, and poor social communication skills. Some have very challenging behaviors, like running out on the street full of traffic, turning on the hot water tap, and screaming for no reason. The range is huge. Some kids manage to go to normal schools with extra support. Some have a very successful life that you hear about now and then in the media. But those at the other end, you may never know. Some are in special schools where, if you enter for the first time, you feel like you are walking into another world. Families won't talk about it. But deep inside, you cannot help wondering. You try to put yourself in those kids' families and try to guess how they feel. It's hard. Still, it's incredible to see those parents, teachers, therapists, and professionals devote themselves to making some difference, regardless of how small, in those kids' lives.

Each time working with a different kid, I learn something new. Like Emily (name changed), a beautiful 14-year-old autistic girl who can only say fewer than 20 words, whenever she has her period, she smears it on the wall, screaming. The school has to arrange for another nurse to support her one-on-one on those days. Despite that, sometimes it gets too much, and the school has to send her home.

Things like these take a toll on the family. And you can see it on the parents' faces. One day, while waiting in a dental clinic for an assignment, I noticed a middle-aged woman sitting opposite me. She looked fazed, weary, and distant. I had a curious Sherlock Holmes moment and tried to guess – what could be wrong with her? Maybe she has a kid with autism? Three days later, by weird coincidence, I did a home visit job to a new client and met her. It turned out I was right. She had a nine-year-old girl with autism who constantly screamed and rarely slept at night. As a single mum, she had to supervise her almost 24×7. When asked what support she most needed, she said, "I need sleep."

But the worst case probably was Nole (name changed), a 17-year-old boy with autism. One day, I walked into a special development school for a parent-teacher meeting. The teacher showed me the classroom for senior students. While waiting for Nole's mum, I briefly chatted with his new Canadian teacher about her life in Australia, this new school, and students with severe disability. While casually looking around, I noticed a bookcase shelf, not filled with books, but nappies. Lots of nappies. Whom for, I wondered. The teacher noticed my stare, softly saying, "For Nole."

Driving home that night, I couldn't help but think about the boy and questions I'm often puzzled about right and wrong, high and low, where we are in this universe, and how far we can reach.

Life has a pattern. If you recognize one, you can predict the next. At least, it's a theory. Working with those disadvantaged kids, sometimes I wonder if we had an intellectual disability, how we would know. How would we know if we cannot teach, rather than they cannot learn? How would we know if we don't understand well enough to explain, rather than they're not smart enough to understand? How would we know what we know is right, if our mind is wrong? We wouldn't. We only have five senses and a brain. Aliens with super IQs out there might have ten senses plus two brains. They would be smarter than us, and we would be just kids to them. They could read our minds, edit our future, and teleport through time. It sounds scary, but I trust the law of intelligence. The stupid would try to make you stupid, but the smart will always make you smarter. So don't worry about meeting new horizons wherever you are. Seize the moment, expand your life.

I get up and step into another world.

# CHAPTER 1
# WAR ON RELATIVITY

*The greatest enemy of knowledge is not ignorance,*
*it is the illusion of knowledge.*
**STEPHEN HAWKING**

*I concluded that the theory is not a theory at all,*
*but simply a number of contradictory assumptions*
*together with actual mistakes.*
**LOUIS ESSEN**
*Physicist and creator of atomic clocks, on Relativity*

*I still get two or three letters a week telling me Einstein was wrong.*
*Nevertheless, the theory of relativity is now completely accepted*
*by the scientific community,*
*and its predictions have been verified in countless applications.*
**STEPHEN HAWKING**

*People express astonishment that Einstein was not awarded*
*the Nobel Prize for Relativity, considered by many*
*to be one of the most outstanding achievements of this century.*
*I do not hesitate to declare that it is not only among the most*
*sensational fancies, but is also one of the most serious logical*
*incoherencies in the history of Science.*
**HARALD NORDENSON**
*Former Nobel Prize judge*

*I think scientists should be humble*
*and not think they're the masters of today's knowledge.*
**CARLO ROVELLI**

Einstein is my hero. I did a lot of portrait drawings when I was a kid, and he was my first. Next was Newton, Leibniz, Euler, Beethoven, Chopin, Van Gogh, Picasso and more. I had no idea why I was fascinated by those people. I doubt if it was their fame. But maybe something that we call the beauty they had brought to life.

Relativity is the first thing I heard about Einstein. And time dilation is the only thing I'm curious about Relativity.

I wouldn't be so fascinated if Relativity was not about time – something so familiar everyone knows, uses, and talks about every day. We dream about time. Kids dream of growing up fast. Adults dream of staying young forever. Elders dream of turning back time. That's why when we first hear about Relativity and time dilation, though we know nothing about science, we're curious to know why and how.

I first studied Special Relativity in Year 12. It was a chaotic time after the end of the Vietnam war, and we didn't have many textbooks, only some lecture notes. Interestingly enough, I learned that speed can slow time. If you travel on a high-speed trip, you will come back younger than your twin brother at home, according to Einstein's Twin Paradox. Apart from that, Special Relativity was quite easy to pass in high school; you just follow its equations and formulas.

And that's what my best friend failed. He was the smartest kid in my class, but never believed in time dilation. He refuted Einstein's calculation. He always challenged Relativity, and I always defended it. Not because I understood, but because I trusted probability. The theory has been tested for a century, approved by all scientists, and learned by millions of university students worldwide. They would never teach a wrong theory in high school, so what are the odds for a student to prove that Special Relativity is wrong? Absolute zero.

Only many years later, after migrating to Australia, there came a time when suddenly I realized I was free – to think. I thought about my childhood, my best friend with whom I'd shared so many fiery arguments about Relativity, and I wondered the impossible – what if the kid was right? It was the first time I wanted to study Relativity, not to score, not to pass, not to win, but just to understand.

I started to read more, some 50 books, all pro-Relativity. The problem was, the more I read, the more I leaned toward my friend, and the more skeptical I felt against Special Relativity. Nothing from those books could I find was convincing. It had reached the point where I struggled to trust those authors. I respected them all. They were all great scientists who wanted to communicate with readers about things they felt passionate about. But I doubted if they saw the picture behind the numbers. Some avoided key questions about time. Some just repeated Einstein's exact words and offered no better clarification. Some accidentally gave contradictory interpretations of Special Relativity to their colleagues, though they all supported Special Relativity. Is Relativity relative? Fed up with unanswered questions, I stumbled to the other side, looking for an alternative.

I found another world. A world of opposition, criticism, and rejection. Contrary to what I'd been fed to believe, Special Relativity is not fully understood, tested, or accepted. Many scientists still oppose it. Dissident physicists are still against it. Evidence is not evidence, many still claim. And their criticisms make sense.

It's a war on Relativity. (See Figure 1.1.)

Gerhard Kraus published a book titled *Has Hawking Erred?* in 1993, claiming Einstein's theory is a major fallacy. He wrote:

*For a long time I had harboured doubts about certain mind-boggling views held by Einstein and his modern successors: that time can slow down, come to a standstill and even turn backward; that three dimensional voluminal space can accommodate additional space dimensions, one of them being time, resulting in space–time; and that voluminal space and time can warp and bend.*

*Figure 1.1 "War on Relativity" and the Australis Ballet Dancers*

*While appreciating Hawking's book (A Brief History of Time) and his other admirable contributions to modern physics, I believe that I have found several specific contradictions in his text space–time concept. This concept was adopted by Einstein several decades ago and has since become something of a sacred cow to most physicists the world over. Despite such universal acclaim, I have become convinced that by uncritically accepting Einstein's space–time theories, which play such a large part in theoretical physics, Hawking, in common with other physicists, may be perpetuating a major fallacy unique in scientific history.*

*What I like to emphasize in this respect is that my approach to Hawking and Einstein is motivated by a healthy skepticism. I am merely taking a leaf from Einstein himself, who, according to Barnet, proclaimed that he was unwilling ever to accept any scientific principle as self-evident, merely at its face value. To me the principle of space–time, as it is postulated by Hawking and Einstein, is far from self-evident.*

I had the privilege of communicating with Kraus when he was in Thailand. It was through him that I learned more about skepticism against Special Relativity from the general public.

Even Professor Paul Davies wrote in his book *About Time* in 1995:

*Among the usual stack of letters, circulars and memoranda are three fat manuscripts sent to me from private addresses: England, California and Western Australia. All came unsolicited and accompanied by letters that start out in the same vein: "Although I am not a scientist ..." I skim the pages of these manuscripts warily. Like many colleagues, I receive several of them each month. Today they are similar in style and content. Two have some mathematics, handwritten, at pre-highschool level. The message is the same: "Einstein got it wrong; I've got it right. Please help me to tell the world." Close scrutiny reveals the authors' deep anxieties about time. How can something so basic to our experience be relative? They protest. That would surely lead to paradox. Something must be wrong. The manuscripts contain complicated diagrams showing observers whizzing about with clocks, and agonized questions about whose time is right and who is being misled.*

Of course, Professor Davies supports Einstein's theory and tries to explain it. But the only answer I got after reading his book is the same statement he wrote in the Preface:

*Nevertheless, you may well be even more confused about time after reading this book than you were before. That's all right; I was more confused myself after writing it.*

But that's not all right. If Professor Davies cannot understand it, who else can? If a well-established academic physicist writing a book to explain why Einstein has completely changed our understanding of time and space, is still confused about time, then how about us?

After all, it's still a great book written by a courageous mainstream physicist who dares to tackle time dilation that many others would avoid. You don't really know problems until you face them. His humble confession exposes the awful truth. Time dilation only exists for those who cannot define what time is.

A similar thing happened to Professor Herbert Dingle, President of the Royal Astronomical Society and Professor Emeritus at London's University College, who initially taught and wrote textbooks supporting Relativity. But in his later years, after constantly facing questions from stubborn students, he became skeptical about the logic of Einstein's twin paradox and completely turned against the theory. In 1972, he published his anti-relativity book *Science at the Crossroads*, in which he accused the scientific community as *"A conscious departure from rectitude."*

He also made an outrageous comment about Albert Einstein and Special Relativity:

> *It is impossible to believe that men with the intelligence and ability to achieve the near miracles of modern technology could be so stupid, so lacking in critical self-awareness, as to accept the relativist doctrine without, so far as I can see, any logical or empirical foundation whatsoever.*

> *How is it possible that such an obvious absurdity should not only have ever been believed but should have been maintained and made the basis of almost the whole of modern physics for more than half a century?*

After retirement, he still faced fierce opposition from the physics establishment and was labeled a "cantankerous philosopher" for criticizing such a monumental revolutionary theory in physics.

Such a monumental physics theory? Ever wonder why it was not awarded the Nobel Prize?

Here is the reply from the former Nobel Prize judge, Harald Nordenson:

*People express astonishment that Einstein was not awarded the Nobel prize for Relativity, considered by many to be one of the most outstanding achievements of this century. I do not hesitate to declare that it is not only among the most sensational fancies, but is also one of the most serious logical incoherencies in the history of Science.*

Nordenson also published a book criticizing Special Relativity, titled *Relativity, Time and Reality*. He wrote:

*The result is that Einstein's Theory of Relativity is based on the indiscriminate use of the word "time" in two different meanings which makes his Theory untenable from a logical point of view.*

*The Theory of Relativity is not physics but philosophy and in my opinion poor philosophy.*

We're not only talking about early criticisms soon after 1905, but almost a century after Einstein's first publication, how strangely people within the physics community are still trying to challenge it.

*The Scientist* magazine, May 15, 1995, mentioned a group of dissident scientists challenging Einstein's Special Relativity. It wrote:

*In Norman, Okla., this month, about two dozen speakers will gather to challenge dominant paradigms of modern theoretical physics and to discuss alternatives. At the annual meeting... these self-styled dissidents are planning to renew their attack on the special theory of Relativity and big-bang cosmology.*

*They claim that they cannot publish their ideas in mainstream journals and that the bias against them is so extreme that, in essence, one cannot become a physicist if one criticizes the special theory of Relativity. Indeed, the one exception to the rule – a physics professor at the University of Connecticut – did not reveal his skepticism of Albert Einstein's theory until after he had tenure.*

John D. Chappell, chairman of then Natural Philosophy Alliance, confirmed the same in *The Scientist*, March 18, 1996, when his professional role was questioned:

*It is incorrect that very few of us are physicists. Nearly all of us are: some work as physicists or engineers in industry, some teach*

7

*in other disciplines, and some have been forced into unrelated work. Nearly all are excluded from academic physics by a strict weeding out of Special Relativity (SR) critics starting in the undergraduate years. A physicist once threatened to destroy my Ph.D. program outside physics if I criticized SR in my thesis. Since 1960 there has been virtually total censorship of criticisms of SR in mainline physics journals, and in planning major sessions at national physics and general science meetings.*

Edwin F. Taylor, a physics lecturer at Boston University, replied also in *The Scientist* magazine, March 18, 1996:

*As former editor of a physics education journal and coauthor of two Relativity texts, I have had a lot of contact with what your article calls "dissident" scientists, many of whom attack Special Relativity. In my experience, most are extremely intelligent and inventive. Their arguments are ingenious, and any errors are often difficult to find.*

*It is heartening that the dissidents are apparently developing a group identity. Your article leaves the impression that at present this cooperation is largely political, a united effort to obtain public hearing for their ideas. If this can be extended to an internal dialogue about the theories themselves, everyone will benefit.*

*Right or wrong, the published theory could provide a monthly challenge to young scientists to question or validate what they are being taught by their elders.*

Books, journals, or articles criticizing Einstein's Relativity are not easy to publish. Once published, very few general readers would ever hear of them. Here are a few:

"Bashing Black Holes: Theorists twist Relativity to eradicate an astronomical anomaly" by J. Horgan, in *Scientific American*, July 1995; "The Yilmaz challenge to general Relativity" by F.I. Cooperstock and D.N. Vollick in *Nuovo Cimento* 111B (1996); *Unsolved Problems in Special and General Relativity*, a collection of 21 papers edited by Florentin Smarandache, Fu Yuhua, Zhao Fengjuan, and many other critics from China, Russia, Japan, and Romania (2013); *Einstein Plus Two* (Golem Press, PO Box 1342, Boulder CO 80306) by Petr Beckmann in 1987, briefly reviews the various relevant experiments and explains why they don't quite prove Relativity.

They don't quite prove Relativity? No, many of those tests thought to prove time dilation are also now challenged to disprove it. The Hafele-Keating test with flying atomic clocks around the Earth in 1972 is a popular one.

Dr. Al Kelly, the author of *A New Theory on the Behavior of Light*, 1995, and *Challenging Modern Physics*, 2005, questions many issues in Einstein's Relativity theories. When Al Kelly obtained Hafele and Keating's 1972 original test data from the US Naval Observatory, he discovered *"extensive undisclosed alterations."* Kelly wrote:

> *Tests that purported to confirm the requirement of Special Relativity, that moving macroscopic clocks run slow, were carried out by Hafele & Keating by flying atomic clocks in opposite directions around the Earth. These tests have been shown to be seriously flawed and to provide no such evidence.*

> *The Hafele & Keating experiment may well rate as one of the biggest hoaxes in the history of modern Science.*

Even Hafele, in his analysis published in *Nature*, stated that:

> *The standard answer — that moving clocks run slow — is almost certainly incorrect. The difference between theory and measurement is disturbing. Most people (myself included) would be reluctant to agree that the time gained by any one of these clocks is indicative of anything.*

More than just the Hafele-Keating test, Kelly's book is almost an encyclopedia about Relativity and its problems, such as the Sagnac effect, the twin paradox, gravitation, and so on, and even his own "Universal Relativity." After a brief dialogue with him, I was blown away by his vast knowledge, extensive research, and deep passion for the subject. He's a distinguished engineer with an immaculate record of community service, not only in Ireland but many third-world countries.

No wonder his controversial book has caused such a media frenzy. And, of course, counter criticism, such as this from physicist John Farrell (May, 2011):

*It was not the work itself that shocked me. Any physicist regularly receives refutations of SR in the post, ranging from the completely crazy to the highly aggressive. What shocked me was the media reaction to Kelly's work; the story immediately became a media "controversy", with feature articles in The Irish Times (Ireland's paper of record), the Sunday Times (a respected UK paper), the Japan Times and many others.*

*All in all, the affair left me with a very skeptical view of science in the media, a view which has not changed much over the years. It seems to me that there are several problems in science journalism that are rarely aired; – motivation: the primary motivation of a journalist is to get the story, not the truth; a difference of opinion quickly becomes a "controversy" – expert opinion: journalists can be quite hazy on what constitutes expert opinion; the Kelly story would only have been controversial if he was a Professor of relativity (or even a physicist)*

*– scientific knowledge: journalists often have quite a low level of understanding of the science in question, and even of the scientific method. In particular, a great many journalists fail to grasp the difference between discovery (initial theory etc) and justification (evidence)*

*– scientific consensus: this remains a mystery to many journalists; how it is achieved and why it is important yet never unanimous*

*As a result, I have become convinced that there is a need for a new sort of science journalism – science communicated by scientists, not by journalists.*

Farrell's last comment is interesting – about communication issues between scientists and the public. Should scientists have more training in speaking the common language that the public use? Otherwise, who else, if not journalists, can make scientists understandable? Physics establishment?

Here is part of a joint letter from the Physics Department, Trinity College and University College, Ireland (by Dr. C. O'Raifeartaigh and Dr. L. Hanlon – March, 1996):

*As members of the Irish physics community, we are mystified by the intense interest shown by the Irish media (including The Irish Times) in the writings of engineer, Dr A. Kelly, on the subject of relativity.*

*Comment from established experts in the field has not been forthcoming, perhaps due to the fact that Dr Kelly's "theory" has not been published in any recognised journal concerned with this subject and therefore does not merit their consideration.*

*Two points should be clarified:*

*(1) To report that Dr Kelly gave a lecture in Trinity College (The Irish Times, February 16th), without pointing out that he was invited by undergraduate students to give a talk to a student society, creates an entirely misleading impression of acceptance of his work on the part of the academic staff.*

*(2) The publication of scientific work in internationally recognised journals through a process & peer review is the tried and tested method of introducing bona fide work into a forum for informed scientific debate. Until the work of Dr Kelly has been published in a journal of repute in this field, it must be regarded with the utmost scepticism (remember cold fusion!). Public interest in science is always welcome; however the active promotion in the media of a "relativity theory" which has not been recognised by professionals in the field does not mirror the cautious, objective approach of scientists and does little for the reputation of Irish physics abroad.*

It's a great effort from the physics establishment to make a connection. But what can we learn from that announcement? Does it answer any questions raised in Kelly's book? Or the only thing they announce is, "Trust me, I'm a doctor"? What if the patient no longer believes his doctor and wants a second opinion?

This is a second opinion from Dr. Louis Essen, English physicist and head of the National Physical Laboratory, who invented the first atomic clock. Essen has published a booklet titled *The Special Theory of Relativity: A Critical Analysis*, in which he carefully examined all the problems created by Special Relativity, especially its thought experiments. Using plain language that everyone can understand, he outlined his criticism against Special Relativity:

*Perhaps the strangest feature of all, and the most unfortunate to the development of science, is the use of the thought-experiment. The expression itself is a contradiction in terms, since an experiment is a search for new knowledge that cannot be confirmed, although it might be predicted, by a process of logical thought.*

*The contraction of length and the dilation of time can now be understood as representing the changes that have to be made to make the results of measurement consistent. There is no question here of a physical theory but simply of a new system of units in which c is constant, and length and time do not have constant units but have units that vary with $v^2/c^2$. Thus they are no longer independent, and space and time are intermixed by definition and not as a result of some peculiar property of nature.*

*If the theory of relativity is regarded simply as a new system of units it can be made consistent but it serves no useful purpose.*

*A critical examination of Einstein's papers reveals that in the course of thought-experiments he makes implicit assumptions that are additional and contrary to his two initial principles. The initial postulates of relativity and the constancy of the velocity of light lead directly to length contraction and time dilation simply as new units of measurements, and in several places Einstein gives support to this view by making his observers adjust their clocks. More usually, and this constitutes the second set of assumptions, he regards the changes as being observed effects, even when the units are not deliberately changed. This implies that there is some physical effect even if it is not understood or described. The results are symmetrical to observers in relative motion; and such can only be an effect in the process of the transmission of the signals. The third assumption is that the clocks and lengths actually change. In this case the relativity postulate can no longer hold.*

In a letter to his friend Dr. Carl Zapffe in 1984, Essen briefly confirmed his argument and its rejection from mainstream publishing houses:

*The clock paradox, for example, follows from a very obvious mistake in a thought experiment (in spite of the nonsense written by relativists, Einstein had no idea of the units and disciplines of measurement). There is really no more to be said about the paradox, but many thousands of words have been written nevertheless. In my view, these tend to confuse the issue.*

*One aspect of this subject which you have not dealt with is the accuracy and reliability of the experiments claimed to support the theory. The effects are on the border line of what can be measured. The authors tend to get the result required by the manipulation and selection of results. This was so with Eddington's eclipse experiment, and also in the more resent results of Hafele and Keating with atomic clocks. This result was published in Nature, so I submitted a criticism to them. In spite of the fact that I had more experience with atomic clocks than anyone else, my criticism was rejected. It was later published in the Creation Research Quarterly, vol. 14, 1977, p. 46 ff.*

And his warning for those who criticize Special Relativity:

*No one has attempted to refute my arguments, but I was warned that if I persisted I was likely to spoil my career prospects.*

That warning is real. It has happened to many dissident physicists. It reminds me of Dr. C. O'Raifeartaigh and Dr. L. Hanlon in that joint letter defending Special Relativity. Would they care to read criticisms raised by Essen, Kelly, Nordenson, Krauss, Dingle, and many other dissidents and attempt to refute their arguments? What would possibly happen if they found Einstein wrong? Would they dare to voice their criticism? Would they lose their jobs?

After safely retiring in 1988, Essen wrote another article, "Relativity: joke or swindle?" in *Electronics & Wireless World*, 1988, pp. 126–127, again summarizing his points:

*Einstein's use of a thought experiment, together with his ignorance of experimental techniques, gave a result which fooled himself and generations of scientists.*

*The same criticism applies to a more recent experiment performed, at considerable expense, in 1972. Four atomic clocks were flown round the world and the times recorded by them were compared with the times recorded by similar clocks in Washington.*

*The results obtained from the individual clocks differed by as much as 300 nanoseconds and yet the result was claimed to be accurate to 10 nanoseconds. This absurdly optimistic conclusion was accepted and given wide publicity in the scientific literature and by the media as a confirmation of the clock paradox.*

*All the experiment showed was that the clocks were not sufficiently accurate to detect the small effect predicted.*

*Why have scientists accepted a theory which contains obvious errors and lacks any genuine experimental support? It is a difficult question. But a number of reasons can be suggested.*

*There is first the ambiguous language used by Einstein and the nature of his errors. Units of measurements, though of fundamental importance, are seldom discussed outside specialist circles and the errors in clock comparisons are hidden away in the thought experiments.*

*Then there is the prestige of its advocates. Eddington had the full support of the Royal Astronomical Society, the Royal Society and scientific establishments throughout the world. Taking their cue from scientists, important people in other walks of life referred to it as an outstanding achievement of the human intellect. Another powerful reason for its acceptance was suggested to me by a former president of the Royal Society. He confessed that he did not understand the theory himself, not being an expert in the subject, but he thought it must be right because he had found it so useful.*

*A common physicist's reaction to Relativity is that although he doesn't understand it himself, he thinks it is so widely accepted that it must be correct. Until recently this was my own attitude. But Relativity has always had its critics. Ernest Rutherford called it "a joke". And Frederick Soddy "an arrogant swindle". But today the theory is so rigidly held that young scientists dare not express their doubts.*

*Special Relativity is not a theory, but simply a number of contradictory assumptions together with actual mistakes. I don't think Rutherford would have regarded it as a joke if he had realised how much it would retard the development of Science.*

Frederick Soddy, like Rutherford, was also a Nobel laureate. At a gathering of Nobel Prize winners in June 1954, he made this comment about Relativity: *"A swindle, an orgy of amateurish metaphysics."*

Mainstream physicists would rarely respond to those criticisms. Reasons are plenty. It's hard to find faults in those criticisms. It's tough to argue against critics who know what they're talking about. It's confusing to explain confusing Special Relativity without confusing yourself. It's safe to follow the majority. It's smart to keep a distance. It's wise not to dig deep. And it's easy to ignore. Clifford Will, a pro-Einstein specialist in Relativity at Washington University in St. Louis, made an exception in 1995:

> *Special Relativity has been confirmed by experiment so many times that it borders on crackpot to say there is something wrong with it. Experiments have been done to test Special Relativity explicitly. The world's particle accelerators would not work if Special Relativity wasn't in effect. The global positioning system (which uses satellites to help determine the exact location on Earth of anything possessing a special transmitter) would not work if Special Relativity didn't work the way we thought it did.*

But John Chappell again responded:

> *...the global positioning system provides evidence against Special Relativity and for the existence of an aether.*

He further stated:

> *Most NPA members maintain that no evidence cited on behalf of either theory (Big Bang and Special Relativity) is more than equivocal in meaning, and that the meanings now taught often result from invalid logic and math. SR is only one of many types of relativism that have flourished throughout academia in the 20th century. As those other relativisms imply, the evidence claimed for SR can be interpreted differently, depending upon the viewpoint of the interpreter.*

So what is it? Do we have enough evidence to support Einstein's Special Relativity or not? Have scientists and physicists all agreed about time dilation or not? The answer might be best put by Bryan Wallace, another dissident physicist who couldn't find a publisher and finally had to self-publish his own book, *The Farce Of Physics,* in 1989. He wrote:

> *This book is a journey through my career as a physicist, giving the interesting details of the many events, arguments, evidence encountered along the way. I suspect that the reader will discover the truth can be stranger than fiction.*

*The father of modern physics and astronomy, Galileo Galilei, was outspoken, forceful, sometimes tactless, and he enjoyed debate. He made many powerful enemies, and was eventually tried by the Inquisition and convicted of heresy. In Galileo's time it was heresy to claim there was evidence that the Earth went around the Sun, and in our time it is heresy to argue that there is evidence that the speed of light in space is not constant for all observers, no matter how fast they are moving, as predicted by Prof. Albert Einstein's sacred 1905 Special Relativity Theory. The heresy changes, but as you will find from reading this book, human nature remains the same!*

And his last note to the whole professional physicist community:

*You should realize that in general only about 90% of professional physicists are able to make sense of less than 10% of what other physicists say.*

It's a sad statement from a physicist. It exposes problems caused by Special Relativity, criticisms against Special Relativity, and the physics establishment's failure to face those problems and handle those criticisms.

Another research project, GO Mueller's *95 Years of Criticism of the Special Theory of Relativity (1908–2003)*, with a list of 3789 critical publications, said it all:

*Since 1908 and until this day there exists an uninterrupted tradition of criticism of the Special Theory of Relativity. This criticism has been successfully suppressed and excluded from scientific discourse since about 1922, more or less in all countries.*

*The suppressed criticism of some 80 years has never been discussed in science and therefore has never been refuted. As a logical consequence, the Special Theory of Relativity remains an unconfirmed hypothesis and has never reached the proclaimed status of the "best confirmed" theory of physics. The many suggested "experimental confirmations" are refuted by the critics as wishful thinking if not accused to be a deception of the public.*

*We have started a Research Project of international scope based somewhere in Germany which documents all publications critical of the special theory of relativity (including some criticism of general relativity) in all languages and from all countries having been published until 2003 and up to the present.*

*The main reason for this tragic success of a new type of "science" we see in the fact that the academics have managed to hide the expulsion of the critics from the public. The public in all countries has been and is cheated about this fact to the present day.*

*The 1922 created new type of "science" we call Socio-Physics. How does Socio-Physics work? The academics and the media tell the public that there is no criticism of relativity. There have been only a few criticisms in the early years after 1905 but it has been "refuted". If there should be expressed any criticism of relativity it comes from cranks and crackpots who do not deserve any attention. No serious scientist has any doubts about special relativity. Therefore the journals have no critical papers to publish, the international conferences have no critical contributions to listen to, the academics have no critical discussions, the scientific publishing houses have no critical books to publish. For experiments which may contradict relativity there is no money left. Socio-Physics has organized a complete suppression of criticism and has everything under control.*

*We have to face the reality that the theoretical physics has been organized as a conspiracy against the public. If our diagnosis of the Relativity Catastrophe and the resulting Socio-Physics is correct, we have to find ways to inform the public about this state of affairs.*

What is the purpose of that project? What is the purpose of all those criticisms? Professor Lutz Kayser, a rocket pioneer, offered an answer in his book *Falsification of Einstein Theories of Relativity* (2015):

*For the sake of future students and scientific truth we implore the physics community to recognize these refutations and start a thorough damage repair. I know that this needs lots of soul searching and changes in text books and attitudes. But the Ptolemean system had to be changed, too, against ecclesial doctrine. Unfortunately, the problem is of similar magnitude, since Einsteinian Theories have become like a religion for their believers. Students and researchers of physics, engineering, and even medicine have flunked their exams, lost their PhD grade, or even were denied professorships because they doubted Einstein's Relativity Theories.*

The same sentiment was earlier shared by many, such as Eugene Mallove, Sc.D., who wrote this in "Breaking Through Editorial: The Einstein Myths – Of Space, Time, and Aether" (Originally Published July-August, 2001 in *Infinite Energy* magazine, issue 38):

> *Our central objective is to show that such criticism does exist, that it is reasoned, and that there have long been open questions about relativity, which have been deliberately ignored by the Physics Establishment. We hope that this coverage will inspire those who remain free-thinking and who are not intimidated by the prevailing intellectual tyranny that passes for physics today. We hope especially to reach the uncorrupted – young students of physics who may help pioneer new ways of experiment and understanding.*

Can you believe all that?

I was shocked when I first read those lines. Are those quotes real? Where are the sources? What if they're all fake? You shouldn't believe whatever you read these days, should you? I don't. My concern is never about the sources, only the target. People can write anything they want; only you can decide what you think. Regardless of whether they are real or fake, do they make sense to me?

They do. They confirm my doubts. They resonate with the exact questions my friend raised in high school.

I hated physics and never wished to be a scientist. Despite having a degree, my highest physics knowledge is probably at junior high school level. It was my friend who cared about this and deserved to understand Relativity. Not me. I only want to understand it because I want to understand my friend.

There is a need for laypersons to understand experts without becoming one.

If experts cannot convince the public, what are experts for? If sellers cannot sell what buyers want to buy, what do they sell? If the smart cannot convince the foolish, why do they call themselves smart? If physicists cannot convince their colleagues, why do they teach students in schools?

I need answers from a genius who can convince me and speak my language.

Imagine if you could meet one in a dream. Who would that be?

Who else, but Sir Isaac Newton?

# CHAPTER 2
# INTERVIEW WITH NEWTON (Part 2)

*Imagination is more important than knowledge.*
*ALBERT EINSTEIN*

*Figure 2.1 Interview with Newton*

I met Newton again one day in winter at his residence. He wore the same dress the last time we met. Though looking a bit more tired, he still had the same authority, arrogance, and anger when we talked about the downfall of space and time in his world. (See Figure 2.1.)

"Why are you here?" he finally asked.

"I'm confused, Sir," I said.

"Of course, you are. So is this stupid world."

"Can you help?"

"About what?"

"Einstein's Relativity. Space and time."

"Didn't I explain everything last time?"

"You did. I'm very thankful for that — it helped me understand more about problems with Special Relativity. But the world hasn't changed. For over 100 years, the public has believed in Special Relativity. The same childish arguments and evidence. I have no idea how they could convince anyone. Yet millions of students worldwide still study and accept the theory. No one seems to know about anti-Relativity arguments, which make a lot of sense to me. I asked some mainstream physicists about those, and some say they've never heard of them, while some say they don't care and regard those critics as idiots, cranks, or crackpots."

"Why do you think this is?"

"There's no reward for challenging Relativity. Mainstream physicists are too busy with other work. They have no time to listen to those they regard as 'crackpots.' They're comfortable with what they know as the truth."

"Then why does it bother you?" asked Newton.

"Because I don't believe that truth."

"What do you want from me?"

"Does Einstein's time dilation bother you?" I asked.

"Of course it does."

"Then why don't you do something about it?"

"You mean to stand up and declare that all those scientists are stupid?"

"No. But why don't you at least try to convince them? They're your colleagues. If one scientist can convince all other scientists, it must be you. Don't you agree?"

"Let me tell you something, son. First, they don't need me. Relativity is not hard enough for them to need me. They're smart enough to find truth for themselves if they want to. But again, as you've said, they're already comfortable with truth they have. There's no reward for challenging it. Second, I'm almost four centuries old. Technically, I'm already dead. My time's expired. Your world needs me no more. I might sound sarcastic, but I believe in young scientists. It's their time. Let them shine."

"But they cannot graduate if they challenge Einstein."

"Sounds like an education problem, not a knowledge one."

"Aren't they the same?"

"Not quite. But tell me, what is the purpose of education in your world?" asked Newton.

"To teach knowledge," I replied.

"How?"

"First, they teach you what to say; then they test what you remember. If your answers match their answers, they pass you. If not, they fail you."

"What's wrong with that?" asked Newton.

"Nothing. It is what it is. Education is about copying true knowledge. There should be no shame in doing that. If you cannot copy knowledge, what's knowledge for?"

"What if you copy wrong knowledge?"

"Yeah, then it's a problem. No one wants to teach or learn wrong knowledge. But how do we know if our knowledge is wrong? How do we know what we know is true is really true?"

"Keep asking. Keep searching. One day you'll find the truth. Answers won't come unless you ask. It's what education should be – not only to pass on knowledge but also to test knowledge by questioning. Wrong knowledge can pass through education if students are not free to ask questions. Einstein was not questioned enough by other professors. Professors were not questioned enough by teachers. And teachers were not questioned enough by students. That's why we end up with this war on Relativity where only graduates who agree pass and graduates who disagree fail."

"*You* can make a difference, Sir."

Newton wasn't impressed.

"State your request."

"The world needs you. We want to hear your story. If we listen to only one side, we only have half – not the whole story. I know you're sick and tired of trying to convince people. Some don't deserve that. But some do. Imagine somewhere out there, a 17-year-old schoolboy has been searching all his life for you. He loves math, physics, and science. He loves Einstein and Newton. But most of all, he loves truth. He's not convinced by the truth presented by all pro-Einstein and pro-Relativity arguments. He finds that truth is not true. He left the other side because he no longer believed. He left because people over there treated him as retarded. Now he's here because he wants to listen to you. He wants to test the wrong side of his brain, to reach the wrong side of the universe, and to see why right things don't work. Can you please give him a second chance in life? Can you please teach him a lesson? Can you please show him questions that answer everything?"

Newton sighed, shaking his head. It was not a good sign. Then came the question.

"What do I get in return?"

I was stuck. I had no idea. But seeing some Macca's wrapping paper on his table, I took a punt. "How about… a Big Mac meal each time we talk?"

He frowned, then replied after four seconds.

"Deal."

# CHAPTER 3
# MAGIC SALON

*To know everything is to know nothing,*
*but to know nothing is to know everything.*
*CONFUCIUS*

*If knowledge can create problems,*
*it is not through ignorance that we can solve them.*
*ISAAC ASIMOV*

*If you can't explain it simply, you don't understand it well enough.*
*ALBERT EINSTEIN*

"What do you know about Special Relativity?" Newton asked me the next day we met.

"Why?"

"You said you've read over 50 pro-Relativity books, right?"

"Yes."

"Do they make sense to you?"

"Not really."

"Then why do you read them?"

"I'm curious. I try to find the truth. And I can't find it."

"From what you've read, what is Special Relativity? Tell me in less than 100 words."

"You want me to summarize 50 books in less than 100 words?"

"If you can't summarize, you don't understand."

"I can summarize. I still don't understand."

"You might not understand why they say it, but first, you must know what they say. What do they say?"

"They say that Einstein's Special Relativity consists of two postulates. First, the laws of physics remain the same in all inertial references. Second, the light speed (in a vacuum) is constant, independent of the speed of the source and all observers' movement. In other words, regardless of the speed you travel relative to a light source, you would always measure the light speed as the same. As a result, when an object moves, its time slows down, and its length contracts. Another one is the twin paradox, where a twin traveling in a high-speed spaceship upon return will find himself younger than his twin remaining on Earth. Numerous tests (including some using high-accuracy atomic clocks) have all confirmed the effect of time dilation. Despite its anti-intuitive, mind-boggling views, Einstein's Special Relativity has completely changed the way we see our universe: only light speed is absolute; everything else, including simultaneity, is relative. And time is simply another dimension in his revolutionary 4D space–time model. Am I over 100 words?"

"You are."

"Sorry. But is it a fair view about Special Relativity?"

"It's too long. You talk too much. And so does your world. Too many Special Relativity books talk too much about irrelevant things, and that's why no one understands. I can sum up the whole of Special Relativity in just 10 words."

"How?"

"SR: {inertial motion $\rightarrow$ time dilation, length contraction}

Or even shorter:

SR: {v $\rightarrow$ (t < t'; L > L')}"

"Cool."

"And you said you don't understand Special Relativity?"

"No."

"Or you do, but you find it's all wrong?"

"I guess it's the same thing. When you hear something wrong, which is said and confirmed by so many experts, you have to say sorry, I don't understand. It's just a humble way to put it."

"I don't like your kind of cheap humble sweet talk. I'll tell you straight what I think. Special Relativity is all wrong."

"Why?"

"Because the second postulate is wrong."

"Why?"

"I'll explain later. But first, do you agree with me, if I can prove it wrong, the whole Special Relativity together with time dilation and length contraction will collapse?"

"True. But what about all the evidence, such as the Michelson-Morley tests, Kennedy-Thorndike, Ives-Stilwell, Hafele-Keating, and many more, which all have confirmed Special Relativity?"

"Forget all the evidence. I don't care even if you have evidence from 1000 tests; I can still show you why they've all confirmed something else, not Special Relativity."

"What do they confirm?"

"I want to show you something else first, something more important than evidence – the principle of Special Relativity. Once you understand its fundamental error, the whole theory will collapse – no need to talk about evidence."

"I don't understand."

"Let's just say a man is charged with murder, and prosecutors say they have over 1000 pieces of evidence to prove him guilty. The defense barrister says he doesn't care, and he needs only one. On the day of the hearing, he brings with him the victim – who was thought to be dead but now still is alive. Everyone is shocked. How can you charge someone with murder if the victim is still alive? It's fundamentally wrong. Case dismissed. But now the most embarrassing question, what can you say about 1000 pieces of so-called evidence? Rubbish?"

"What's fundamentally wrong with Special Relativity?"

"To answer your question, I have to tell you three stories."

"No, I don't want to hear any stories. Just tell me—"

"Just shut up and listen; you might learn something," Newton almost shouted. "You've read 50 books, and nothing has convinced you. Why are you here if you can't listen to me? This is what is wrong with Special Relativity and how a whole generation has been fooled. Before you can charge someone with murder, first make sure you have a dead body. If there is no dead body, there is no murder. Similarly, before you can say a theory is right or wrong, first, you must know what the theory is saying. And this is the problem: lots of people, including many pro-Relativity physicists, don't even know what the theory is saying and cannot interpret what it means, let alone have evidence to prove it. What nonsense! What a load of garbage!"

"How is it possible?"

"Are you ready to hear my three stories now?"

"I am, Sir."

"The three stories are: 'Magic Salon,' 'Naïve Fairy Mother,' and '1 = 2.'"

"1 = 2?"

"It's the last story. But listen to my first one first. Imagine a new beauty shop, 'Magic Salon,' just opened in New York, advertising a revolutionary beauty treatment that can help clients become 20 years younger. Dr. X, the shop owner, guarantees immediate results after five one-hour sessions for US$50,000. Many ladies eagerly walk in, but all come out disappointed. Those hour-long sessions are freezing cold, agonizing, and brutal. As a result, they all come out very sick. Some have visibly aged more, and no one looks younger. The only thing they have now is a piece of paper – a new birth certificate issued by that magic salon and signed by Dr. X certifying that they're now all officially 20 years younger." (See Figure 3.1.)

"That is cheating."

"Yes. $50,000 for a fake birth certificate. But the story doesn't end there." Newton continued.

"What next?"

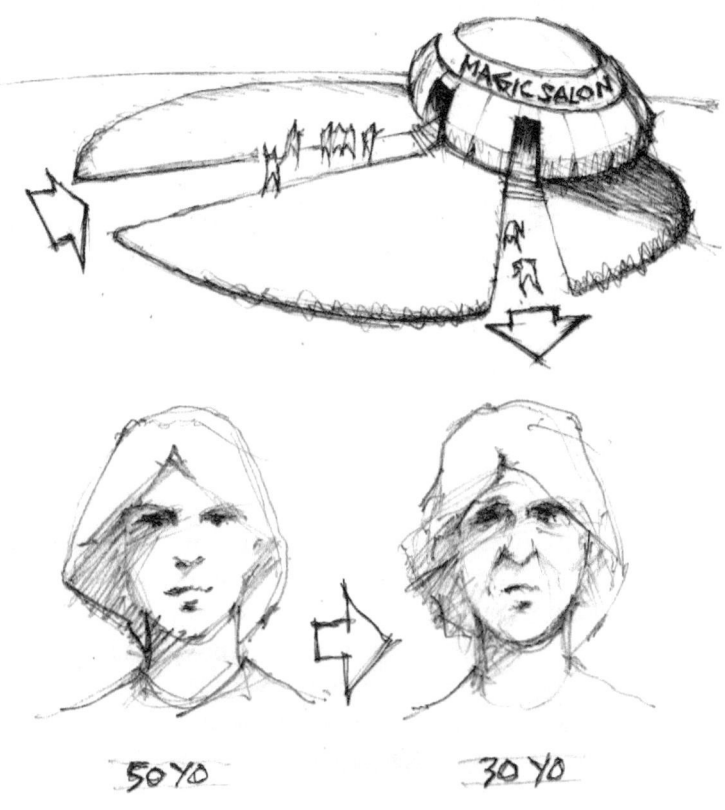

*Figure 3.1 Does changing your DOB make you look younger?*

"As you can guess, all the ladies sue Dr. X for a full refund plus compensation. During the trial, Dr. X explains to the court that he never promised 'younger looks' but only 'younger age.' He exhibits all contracts signed with clients, and it's true that nowhere could you see the word 'younger looks' therein. Dr. X, who is not a doctor of medicine but a specialist in science and mathematics, explains to the court the scientific process behind his revolutionary treatment. He proves to the court that, according to his calculation and measurements, all his clients have obtained a younger age after treatment. His reasoning involves some complex advanced math that many professional witnesses cannot understand, except one math professor who confirms it's all correct theoretically. So scientifically, Dr. X argues that he has done nothing wrong, has fulfilled all his contractual obligations, and blames his clients for their dissatisfaction with the service."

"What does the judge say?"

"The poor judge, who confesses he failed Year 12 math and then switched to legal studies and social justice, tells Dr. X that even if his math is correct, he should keep it to himself or sell it to only mathematicians. However, as a service provider, he has a duty of care to provide a service that customers should reasonably expect. He says customers without math knowledge still have rights fully protected by laws and should never be penalized or tricked into a contract they do not fully understand. The judge is not satisfied that Dr. X has truthfully and reasonably informed customers of the outcome of their contracts; otherwise, no customer would want to pay $50,000 for a piece of paper, regardless of how mathematically correct it is, without achieving any real youthful look. He is satisfied that Dr. X has misled his clients by producing false documents, namely birth certificates. He fines him, orders him to pay compensation, and shuts down his salon."

"Why did you tell me that story?"

"Why do you think I told you that story?"

"Know what you buy?"

"Exactly. Millions of people worldwide believe in Special Relativity because they think the theory offers a scientific solution to forever youth, longevity, and slow aging. That's what they hope. That's what they think Einstein's time dilation means. But does it? Did Einstein ever say that time dilation would mean 'slow aging'? You've read over 50 relativity books. Have you ever read one written by Einstein that specifically states that the twin, coming back from the high-speed trip, apart from obtaining a younger age, would surely *look* younger, have a much

30

*stronger* body, with *firmer skin, fewer wrinkles,* and *less grey hair* than his twin brother?"

"No."

"Did you ever ask?"

"No. Because in general language, when we say 'time dilation,' we mean 'slow aging,' right?"

"Correct. It's what you think. It's what the public thinks. It's what everyone thinks. And it should be that way. But is it the same in physics language or Einstein's language? Or similar to that Magic Salon story, what if Einstein never meant time dilation as slow aging, but just another way of measurement like a new birth certificate and nothing else?"

"We'll be damned."

"Also, what if the opposite happens?"

"What do you mean?"

"What if the twin comes back looking much older?"

"How is that possible?"

"Why not? Imagine locking yourself in a small spaceship for an interstellar trip, lonely, boring, into zero-gravity deep space, far away from Earth, without friends, family, or any social interactions for 20 long years, is it possible that at the end of the trip, (if you survive) you wouldn't look younger according to your new birth certificate, but instead you look much older and weaker than your twin back home?"

"Yes, it's possible."

"Is that why Einstein mentioned nothing about slow aging but only time dilation?"

"Maybe. But what is time dilation then?"

"Good question. Who invented time dilation?"

"Einstein."

"Here is a better question: Why didn't you ask him?"

"I wish he was here. But at least time dilation means clock slowing, right? Experts say that many tests have confirmed it."

"You keep talking about tests and evidence, which I said I would talk about on another day. But since you mention clock slowing, let me ask you which clock are you talking about?"

"Any standard clock, I guess."

"What is a standard clock? Look at the two clocks on my kitchen table. One is blue colored, and the other is red. The blue clock runs a bit slower – about 10 minutes late every six months – and I have to manually correct it every six months. Now, let's say one day, I find a way to fix the blue clock to make it run faster. How would my fixing that blue clock affect the speed of my red clock? If one clock runs slow, why would it affect other clocks?"

"Maybe Einstein meant a standard accurate clock?"

"What is the definition of a standard accurate clock? Do you mean a perfect clock that never slows, never speeds, never breaks, and always shows the correct time regardless of any conditions?"

"Yes."

"Then one day, if that clock runs slow, wouldn't you call it faulty?"

"I don't know."

"Do you think Einstein understood the definition of a standard accurate clock?"

"Maybe he didn't."

"No, he didn't. That's why he was confused about time. And so is everyone in your world. But let's move on for now. Let's just say, yes, time dilation means clock slowing. What does clock slowing have to do with slow aging, then?"

"What do you mean?"

"Let me ask you this: if a man has a heart attack and dies, would his watch stop running?"

"No."

"Then why would a slowing clock affect the rate of his heart, speech, thinking, growth, and aging?"

"It shouldn't."

"Is that why Einstein said nothing about slow aging? Because he was also confused, wasn't he?"

"Yeah, maybe."

"And what about length contraction?"

"What about it?"

"Didn't you say before that according to Special Relativity, when an object moves, its time slows down, and its length contracts?"

"That's what I've read."

"But is it a real length contraction or just a visual distortion during the flight?"

"According to Special Relativity, it's real."

"Are you sure?"

"No, I'm not."

"Let's say we have a 70-meter-diameter flying saucer that normally cannot pass through the Tower Bridge in London, which has a 61-meter clearance in the middle span. The question is: if it travels at 0.8 light speed, resulting in a contracted length, according to Einstein, could it go through the bridge?"

"It should, assuming it's constantly spinning around while flying with its length and width both contracted." (See Figure 3.2.)

*Figure 3.2 Can this flying saucer go through the London Tower Bridge?*

"But how would its width be contracted? Did Einstein say that?"

"No, he didn't. It's only my assumption because of the saucer's spinning motion."

"Let me put it another way. Let's say that during the war with aliens, Earth needs to put a firewall system to detect and destroy any enemy spaceships whose lengths are over 50 meters. The system consists of three 50-meter-wide positive layers separated by two 50-meter-wide negative layers. Any length spanning three different charges (+ − + or − + −) would be detected as over 50 meters long and destroyed. All alien spaceships' lengths are over 60 meters. So there is no way could they go through. But if length contraction does happen according to Special Relativity, could they go through when they flew at the speed of 0.8 c?"

"Yes, they could." (See Figure 3.3.)

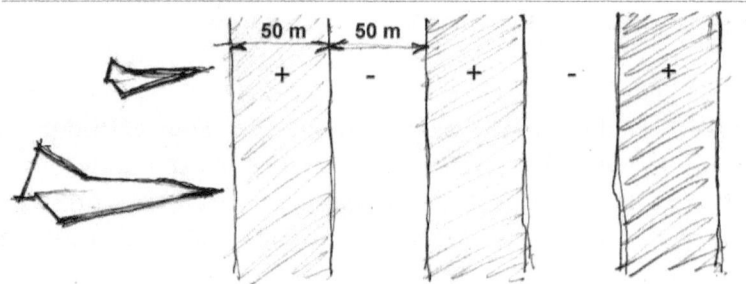

*Figure 3.3 Can alien spaceships go through this firewall system undetected?*

"So, length contraction is real, isn't it?"

"Yeah, I think so, according to Special Relativity. If it's not real, what's the point of having Special Relativity?"

"But if length contraction is real in that sense, does it violate Special Relativity's first postulate claiming the laws of physics remain the same in inertial motion − meaning lengths should remain the same?"

"Maybe Einstein meant the measurement should remain the same because the traveling ruler also got contracted?"

"The length is contracted, but its measurement remains the same. Isn't that a faulty measurement?"

"Yeah, I guess you could say that. That's why I don't understand Special Relativity."

"Let's move on. Talking about length contraction, would length contraction affect the traveler's body as well?"

"I guess so."

"So if you come back years younger than your twin, would your body also become thinner as well?"

"I'm not sure."

"What if you lie down during the trip? Would your body become shorter and fatter instead?"

"Maybe."

"What if you change your position, vertically and horizontally, during the trip? Would your body become shorter in height and width?"

"Could be." (See Figure 3.4.)

*Figure 3.4 Which twin would look younger, older, thinner, or shorter?*

"Let's go back to the previous question. If you lie down during the trip, you said that your body would become shorter and fatter upon return, right?"

"I think so, according to Einstein."

"But during the trip, would you see your body as shorter?"

"You should not. Since the first postulate says that the laws of physics remain the same during inertial motions. If you could see, that means you knew you were traveling, and that'd break the first postulate."

"So, according to the first postulate, you should not see, but why could you not see? I mean, if you become shorter and fatter in one dimension while the other dimension remains the same, surely you could notice that by looking at a mirror, couldn't you?"

"I don't know. Could it be because our eyes become shorter and fatter as well?"

"So you think if our eyes get shorter and fatter, they would visually neutralize any distortion? Is that what optics has taught you?"

"Maybe. I don't know. I can only guess."

"Keep guessing. So let's just say, during the trip, you wouldn't feel any difference at all in your body, and then at the end of the trip, when you step out of the spaceship, suddenly you see your body changed?"

"Yes, that's the problem."

"Let's just assume your body becomes shorter after the trip. How much shorter would it be?"

"Depends on your flying speed, according to Special Relativity's equation."

"Only the speed?"

"Yes, according to Special Relativity's equation."

"So, if you travel for one hour or 10 years at high speed, would you return with the same length contraction?"

"Yes, according to Special Relativity's equation."

"Do you find it absurd?"

"Yes, I do."

"What if you change your speed many times during the trip? What speed would you use to calculate, and what length contraction would you end up having upon return?"

"I don't know. That's why I said I don't understand Special Relativity."

"Is it possible that, like the Magic Salon story, Einstein's time dilation or length contraction is just an illusion, a new way of measuring while traveling, but nothing real at all?"

"Maybe. That's what I don't understand. Actually, among 50 relativity books I've read, some vaguely say the effect is visual and the same between two twins due to uniform motion's symmetry. If A moves relative to B, B can also be seen as moving relative to A. If A measures B's time ticking slower, so would B of A. And that means no physical difference at the end."

"If that is the case, then what's the point of having a theory that makes no physical difference in reality?"

"Exactly. If Special Relativity is another Magic Salon's story – another way of measurement, not a physically real one – then what's it for?" I asked.

"What about other books and other authors?" Newton continued asking.

"Some deny the symmetry. They say the traveling twin has to slow down, do a U-turn, and go back, which breaks the symmetry due to acceleration."

"But what happens to part of the trip where no acceleration is involved? Is that inertial motion?"

"Yes."

"Which one is moving then? A or B?"

"Some books still say it's B."

"Gosh. Which means in inertial motion, we can define which one is moving and which is not?"

"Exactly."

"And that's against the first postulate?"

"Exactly."

"What do they say about length contraction? What does it apply to? The traveling twin, his width only, his height only, his spaceship's length only, or all his dimensions?"

"They say very little. Some say the spaceship's length is measured as contracted during the trip by the home twin. But no one says the spaceship will be shortened upon return. And some say length contraction happens to the journey distance."

"Journey distance?"

"Yes. That means, if the twin travels from New York to Los Angeles, a distance of 2500 miles (4000 km), depending on his speed, he might measure it as only 1000 miles."

"So, according to that author, length contraction is only a measurement contraction, not a physical one, as we know that the NY–LA distance doesn't shrink at all just because someone is flying in the sky?"

"Exactly."

"So, according to that author, likewise, time dilation must also be a 'measurement dilation' or 'calculation dilation,' not a real reading on a clock?"

"I would think so. But strangely, he says real-time dilation does happen to his clock and his age, but real-length contraction does not happen to his body."

"Why?"

"I don't know. That's what he said. It seems that each different author has a different interpretation of what Special Relativity is."

"Tell me about illustrations. Amongst those 50 books, have you seen any illustrations in which a twin *looks* younger upon return?"

"Yes."

"What about any illustrations showing a twin who *looks* thinner or shorter or fatter or smaller upon return?"

"No."

"Why do you think this is? A drawing error or serious confusion?"

"I think it's confusion. Words are vague; people can hide. But drawings are visual; they can expose flaws. That's why many authors avoid illustrations. And when they do, you can see errors, inconsistencies, and contradictions."

"But what about Einstein? Did he ever give us any illustrations? I mean, it doesn't take much to depict someone looking old, or someone looking shorter unless it's not what you mean?"

"He's a thinker. Thinkers don't draw. He used math. Then he interpreted math into words."

"And what if he misinterpreted those math results, or you misinterpreted his words?"

"We're all screwed up."

"Yes, you're all screwed up, son. Special Relativity is a mess. It's all mixed up, confusing, and self-contradicting. Even Einstein could not make sense of time dilation and length contraction. He didn't even know if the effect was real or not, or how to translate his equation into reality, even if it was mathematically correct. That's why he said very little and stopped there. Then let others interpret for him. The problem is that everyone has a different interpretation. We don't know which is which, science or fantasy – Special Relativity or not Special Relativity, Einstein's or not Einstein's. Some say more, some say less, some copycat, and some say nothing. To settle, they do tests. Normally, tests are to confirm what a theory predicts; here, they need test results to figure out what a theory should be predicting. As a member of the public, you should be careful paying for a service from providers who don't even know what Special Relativity can deliver. Before you buy, ask. The more you ask, the more they answer, and soon will you find out if they truly know something or they know nothing."

"But why? Why is Special Relativity a mess? How did all this start?"

"The second postulate – it's wrong. Einstein misinterpreted the Michelson-Morley test result, thinking the light speed was the same independent of the observers' speed. To make numbers add up, he altered the value of time without realizing he'd accidentally violated the most basic principle of measurement: never change units of measurement, or measurements make no sense. Confusion followed. Einstein was confused between time and speed. When a faulty clock's speed changes, he thinks the universe's time has changed. If there is one word, and only one, that has made Special Relativity and will break it, it's 'time.' I'll discuss that in the next story, 'Naive Fairy Godmother.'"

# CHAPTER 4
# NAIVE FAIRY GODMOTHER

*In spite of the nonsense written by relativists,*
*Einstein had no idea of the units and disciplines of measurement.*
*LOUIS ESSEN*

"A long, long time ago, in a rural village lived a very poor peasant," so Newton started his second story the next day.

"The man had no home, no money, and no food left. One day he went into the jungle looking for a fairy godmother who, according to legends, could help make all your wishes come true. After many days of searching, he found the beautiful fairy godmother in a deep cave behind the mountain. The fairy godmother asked what he wanted. The poor man was so hungry after running out of food; all he asked for was some bread.

"The fairy godmother asked, 'What is bread?'

"The poor man checked his bag, found a small piece of bread, and gave it to the fairy godmother. Fairy godmother looked at it, took it inside the cave, and five minutes later returned with many large loaves of bread. The poor man was ecstatic, thanking the fairy godmother, and quickly left the jungle.

"He sold some of his bread, exchanged some for other food, and happily lived for about a week. When his food was low, he decided to go looking for the fairy godmother again.

The second time, the fairy godmother asked what he wanted. This time, instead of bread, he asked for fish.

"The fairy godmother asked, 'What is fish?'

"The man took a fish out of his bag and gave it to the fairy godmother. She then took the fish inside and, five minutes later, came out with a large basket full of fish. She also told the man that he should not expect this kind of favor again due to her busy schedule. The man nodded, gracefully thanked her and left the jungle. (See Figure 4.1.)

"The man sold those fish at the market, built a makeshift house, and bought some chickens. After a few months, though his life had much improved, he constantly thought about the fairy godmother and decided to look for her just one more time. When he met her, he told her this was his last request, and that he would never bother her again.

"The fairy godmother asked what he wanted. He said money.

"She then asked, 'What is money?'

"The man took out from his pocket a hundred-dollar banknote. She took the note inside and five minutes later came out with a basket full of hundred-dollar notes, totaling almost one million dollars. The man was over the moon, thanked her for the last time, and left. With such a huge fortune, the man went home, bought a cattle farm, built a bridge, a clinic, a school, and helped all his villagers. Not only did he change his life, but he also completely changed his community. You can guess what happens next."

"He would marry a beautiful wife and live happily ever after," I responded.

"No. One day, the police came and arrested him."

"Why?"

"All his dollar notes had the same serial number. They were all fake."

"Why?"

*Figure 4.1 Fairy godmother and a poor man*

"That's why the police arrested him to find out. He confessed everything. So the police searched the jungle, arrested the fairy godmother, and charged both of them for handling and producing counterfeit money."

"What happened at the court?"

"The man said he hadn't known that all money notes had the same serial number. He said the money came from a holy godmother, so he sincerely thought they were real. The poor man also confessed that he'd never gone to school and wouldn't know what was defined as real money or what was fake, let alone producing it."

"What about the fairy godmother?"

"The court asked her if she had produced those notes, and she said yes."

"How?"

"She said she'd just cloned them. She had magic hands. She could clone everything. Just give her a sample, and she could reproduce anything the same as the original."

"But money cannot be printed or reproduced, except by government. If everyone could print money, it's fake money by definition."

"Exactly. But she didn't know. She had no intention of breaking the law. She didn't know what money was, what bread was, or what fish was. She said she'd come from heaven, and no one eats or uses money up there. When she saw the poor man, she only wanted to help him by giving him whatever he asked, without knowing why."

"What was the outcome of that court case?"

"The poor man was found guilty of handling counterfeit money, convicted, and had all his money and assets confiscated. He got a one-year jail sentence suspended with a good behavior bond. The fairy godmother was also found guilty of producing counterfeit money and, given her naive and innocent motive, the judge spared her a jail sentence and gave her only a correctional community order for a hundred years."

"A hundred years?"

"She was ordered to do community work helping poor people, like cloning their fruits, cows, pigs, chickens, schools, bridges, and hospitals. The order was originally for only one year, but since her magic service was so great, the judge extended it to a hundred years."

"I see."

"She's still working now somewhere in the world because of that order. If you're lucky enough to meet her one day, you can ask her for anything. But no money, please."

"Why did you tell me this story?"

"Why do you think I told you this story?"

"Know what you sell?"

"Exactly. The fairy godmother is a good lady. Though she sells nothing and only wants to help. However, to help, she must first know what clients need, which is not always what they ask. Without knowing this, she might sell the wrong stuff, causing problems instead of happiness. When the poor man asks for money, he doesn't mean paper but value (or purchasing power) hidden behind it. But the fairy mother misunderstood and gave him valueless paper. Everyone wants to be rich. The richer you are, the more money you have. In that sense, paper money is like the measurement of wealth – a ruler, an index, or an image indicating the level of wealth. If you want to be richer, you have to sell more, create more, or produce more. You cannot be richer by moving the marking on the ruler, changing the index, altering your image, or printing more money. Paper money is a symbol, a measurement, and a representation of value. It's a manmade product solely created for that purpose. Once accepted, it cannot be changed, altered, or reproduced. It's the most basic rule of measurement – you don't change measurement units – otherwise, all measurements make no sense. You're with me?"

"Yes."

Newton continued:

"Like the fairy godmother, Einstein was a decent scientist who wanted to help people by making a difference in life. People came after him looking for a way to longevity, and they asked for a younger age. What they need are the looks, health, and energy associated with that age number, not just the number. But Einstein took it literally. He thought that if he found an equation – a purely mathematical equation – to recalculate or alter that number, the world would be different. So he gave them just that – an equation to produce – a younger age. But it's only a number.

"It would have been a different story had Einstein produced an equation to reduce wrinkles, reduce illness and suffering, or improve health, fitness, and happiness instead.

We would call it, then, an equation of time dilation (even though it's not semantically correct) to his credit. But as an equation of time dilation with time units being altered, it's fundamentally wrong. It makes no sense.

"Time is just a manmade concept to measure our memory distances. As a measurement unit, it must have an absolute constant speed. Once accepted, it cannot be changed, altered, or reproduced. Time cannot be dilated; otherwise, it's fake time. Science is meant to make a difference in life. If you want to change the world, you have to change the world. You cannot change the world by changing its measurement units to give the impression the world has changed. To change measurement units is wrong, illegal, misleading, irresponsible, and dangerous."

"Why is it dangerous?"

"Fake measurements can kill. You can bathe a baby in the water at 38°C, but at 100°C, you can kill her. A truck can take a maximum load of 100 tons, but will collapse at 150 tons. The Tower Bridge, London, has a clearance of 61 meters wide underneath, and anything 100 meters wide would crash into the two towers. A man can hold his breath for up to two minutes, but will be dead after five minutes. A pilot ordered to drop a bomb on his enemies at 1000 meters, and not 900 meters, where his allies are, may face murder charges if he miscalculated and dropped the bomb on the wrong location. An old building will be blown up in three minutes, and everyone needs to get out within three minutes or get killed. On the show *Air Crash Investigation*, you can see thousands of accidents due to faulty measurements, faulty readings, and faulty devices. Stop mucking around with measurement – it means life or death."

"Do you accuse Einstein of using fake measurements?"

"Yes. He violated the law of measurement. He confused everything and fooled everyone by introducing double measurements into Relativity. The same length which is 10 meters on the ground is measured as seven meters by someone flying in the sky! A 10-meter-long spaceship flying in the sky will be measured as seven meters by someone standing on the ground, despite the traveler confirming his spaceship's length still is 10 meters long! People will ask if the length contraction is real, and relativists will say yes, it's real, but the traveler won't detect any difference because his ruler is also contracted. We will wonder what kind of measurement allows a 10-meter-long object to now be contracted to seven meters but still be measured as 10 meters. A faulty ruler? A mismeasurement? Or sheer nonsense?

"One thing can only have one measurement, not two. The purpose of measurement is to stop things from being relative, fuzzy, and confusing. The purpose of measurement is to make things absolute, clear, and certain. One kilometer is one kilometer. Two minutes is two minutes. Three degrees Celsius is three degrees Celcius. By accepting two different measurements for the same thing, Einstein ignored the rule of measurement and created a dangerous situation where we can't tell the difference between real and fake, right and wrong, success and failure. It might look fine, wonderful, and very entertaining for theoretical physicists to debate their fantasy ideas in their fantasy castles when nothing matters. But when things do matter, like when someone could get killed because of that fake measurement, then it's no longer fun. If the same length can be measured as nine meters and 10 meters at the same time, next time, if it shrinks to seven meters, say, a dangerous length, we won't know if it's an actual contraction or just another measurement contraction. Let us never forget that we've had enough disasters due to false readings, false measurements, and false assessments. We need science to stop that, not support that. Science needs to tell us what true measurement to use, and what wrong measurement to avoid. It can't be both."

"So Special Relativity is just a theory of double measurement?"

"Yes, nothing else. By introducing double measurement, or even worse, multi-measurement, Einstein tried to sell Special Relativity, thinking he'd physically changed the world, but instead only created confusion. The same length can be perceived or measured differently by different observers, and the same time-lapse can be perceived or measured differently by different observers. However, everything is still the same, and nothing changes. It's not relativity anymore, but a misapplication of relativity. Scientists shouldn't do that. We should never use relativity as an excuse for wrongness. Don't sell washing machines that cannot wash. Don't sell toasters that cannot toast. Don't sell money that cannot buy. And don't sell time dilation that doesn't come with youthful looks. Science should only sell the real difference people need in life, not just a different name, different label, different cover, different measurement but emptiness inside. Measurement is for describing the world. If we wrongly measure our world, the world we see is no longer real – it's fake. And fake measurement can kill."

"And you said he did all that because of the second postulate – the absolute light speed regardless of observers' movements?"

"Yes. He misinterpreted the result of the Michelson-Morley test and, from there, created a wrong postulate about measurement. I'll talk more about that in the next chapter, called '1 = 2.'"

# CHAPTER 5
## 1 = 2

"Have you ever watched magicians turn lemons into apples?" Newton asked me the next day we met.

"Yeah, many times."

"Do you ever wonder how they do it?"

"Not anymore."

"But you know it's not real?"

"It's not real. It's magic."

"But how do magicians do it?"

"No idea. They must have great hands to hide somewhere in their long sleeves."

"So even though you don't know or can't explain how they do it, you still know it's not real, right?"

"Yeah. It's magic because you don't know. If you know, it's no longer magic."

"People often say that seeing is believing. But here, seeing is not believing. Why?"

"I'm not sure. I hope scientists can explain better. Personally, I think seeing could be deceptive, sometimes. That's why you need better ways to confirm. If it was real, you would see that on the news. Scientists would make a fuss about it. You wouldn't miss it."

"I see. But is it possible for us to make it in the future?"

"Of course, anything is possible. If we could land on the moon, we could land on Mars, and turning lemons into apples could be just a piece of cake in the future."

"But at the moment, when you see a magician turn lemons into apples, you think it's not real."

"It's not."

"What if I tell you there's one scientist who believes that magic act is real."

"Who?"

"Not only that, but he also went on and developed a whole new theory about turning lemons into apples instantly."

"Does it work?"

"We don't know yet. Still a theory."

"Great. If a magic act can inspire a scientist to develop a scientific theory that works and benefits humanity, it's great. But why do you tell me this story? I thought you would tell me about 1 = 2?"

"I'm telling that story now."

"But why did you tell me about the magician?"

"You'll understand after this."

"Okay."

"Have you heard about the math joke of 1 = 2?"

"Oh, so it's a joke. Yeah, a long time ago. It must have been in primary school."

"Still remember how to do it?"

"Can't. But I still remember the answer if it's the same joke."

"Want to hear it again?"

"Yes, please."

| | |
|---|---|
| "Let | $a = b = 1$ |
| Multiply b: | $ab = b^2$ |
| Minus $a^2$: | $ab - a^2 = b^2 - a^2$ |
| Regroup: | $a(b - a) = (b + a)(b - a)$ |
| Simplify: | $a = (b + a)$ |
| Replace value: | $1 = 1 + 1$ |
| Result: | $1 = 2$" |

"I see."

"Know the trick?"

"Yeah. Zero division."

"You're right. Since (b − a) = 0, we cannot simplify the equation using illegal zero division, or we'll have a conflict. Why do you think I told you this story?"

I was stuck. I had no idea. So I did what I always would when I have no idea. I took a punt. "Know what you do?"

"No," Newton grinned. "Know the rule before you play. If you're not happy with the rule, change it. But change before play, not during play. It's the rule."

"I don't understand."

"Did you know Einstein made the same mistake when creating Special Relativity?"

"No."

"Yes, he did – the most basic error in primary school math."

"What did he do?"

"Do you know Einstein's equation of time dilation?"

"Yes. It's $t = t'\sqrt{(1 − v^2/c^2)}$."

"Where did he draw that equation from? Did he draw that from personal experience traveling into the universe?"

"No."

"Did he draw that from observing high-speed trips carrying atomic clocks?"

"No."

"Did he draw that from his calculations?"

"No."

"So, how did he come up with that equation?"

"It's the Lorentz transformation equation."

"Exactly. Einstein used the Lorentz transformation equation to explain why $c + v = c$."

"How?"

"He thought the Michelson-Morley experiment was evidence of $c + v = c$, meaning the light speed was always an absolute constant regardless of observers' motion. That's what he thought. So, he decided to make it the second

51

postulate. But then, he had to figure out a way to make numbers add up. And this is what he did: he altered the value of one second.

"I'll give a simple example – a rough illustration, so you'll get the idea. Let's say c = 300,000 km/s, and v = 30 km/s as Earth's speed around the Sun. Special Relativity's second postulate states:

c + v = c                                                      (?)
300,000 + 30 = 300,000                                         (?)
300,000 km/s + 30 km/s = 300,000 km/s          (?)
300,000 km/1s + 30 km/1s = 300,000 km/0.91s (ok)

Now all numbers add up. It's perfect. Can you see it?"

"Yes, it's correct mathematically."

"Yes, it looks correct outside. But inside, it's all wrong. Illegal zero division leads to 1 = 2; illegal time dilation leads to c + v = c. One second is a measurement unit; you cannot alter time value, or speed measurement makes no sense. It's a contradiction. You must define measurement units first before you can measure. You must define time first before you can define speed. You must define the rules first, before playing the game. Einstein took the other way around: speed first, then time. That's why he was confused. If you use such a shorter second like that, the light speed now can be measured by some observers as 300,000 km/0.91s or 300,030 km/s, and that's contrary to the second postulate. To fix that, he had to introduce length contraction to offset the change. It's all about data manipulation to cover up a false result. It's a mismeasurement of reality."

"It's like moving the goalposts?"

"You're absolutely right."

"So, how do you explain the Michelson-Morley experiment?"

"Let me ask you this. Let's say you have a scale weighing a truck as 300,000 kg. Now you add another weight of 30 kg, but the scale still reads 300,000 kg. How do you explain it?"

"A faulty scale, maybe?"

*Figure 5.1 Two fish might appear as five fish*

"Yes, a faulty scale, maybe. A faulty reading or faulty measurement, maybe. A weight too small for a heavy scale to detect, maybe. The best answer can be the simplest one. No need to create a theory about mass dilation or mass contraction just because of one negative test result. I can give you many similar examples. For instance, If you get lost in a desert and suddenly see a fuzzy oasis from the distance, it could be just an illusion. Two fish in a small aquarium might appear as five fish due to refraction, but it's an illusion. (See Figure 5.1.) A five-meter-deep swimming pool might look only three meters deep when fully loaded with water, but it's an illusion. You take a photo of a car running fast at nighttime, which appears longer in the photo, but is just an illusion. Depending on your camera's aperture and shutter speed, a sprinting athlete could have his feet disappear in the photo, but it's just an illusion. You see a magician turn lemons into oranges; it's an illusion, too.

Similarly, the negative result of the Michelson-Morley test doesn't mean the absolute light speed, but an illusion of absolute light speed. Science has to differentiate illusion from reality. The correct measurement of illusion is still the wrong measurement of reality. Seeing is not always believing."

"You said Einstein had created time dilation by reducing the value of one second, and you think it's illegal. But it's only illegal if you regard time as a measurement unit. Why is time a measurement unit? Could time be something else?"

"What else could it be?"

"I don't know. I'm asking you. What is your definition of time?"

"Let's talk about time."

# CHAPTER 6
# TIME

*You may delay, but time will not.*
*BENJAMIN FRANKLIN*

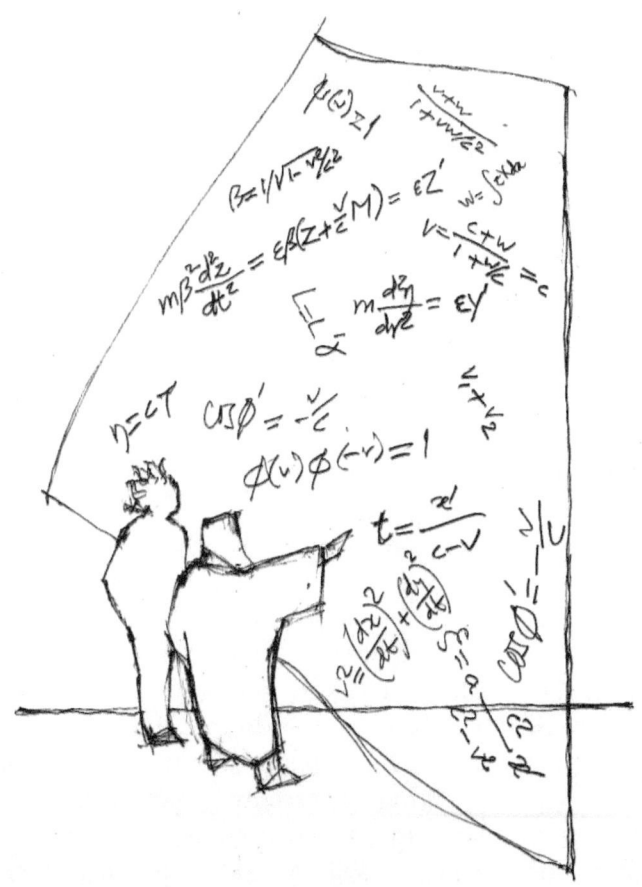

*Figure 6.1 Newton to Einstein: You misspell "t"*

I sat down the next day with Newton and immediately asked, "You regard time as a measurement unit and say Einstein confused time with speed. Why?"

"Before answering that question, I want to ask you this. Do you know much about measurement?"

"No."

"C'mon. We talked about it just last week. Why do we need measurement?"

"You said the purpose of science is to make a difference, save lives, and predict the future. Some differences are so obvious everyone agrees, so nothing further is required. But some other differences are not easy to detect; we need correct measurement to be sure."

"Why is correct measurement so important?"

"Because false measurement can kill."

"Can you give me some examples?"

"You're flying at 1500 km/h, and you think your speed is 500 km/h. You're so close to the mountain ridge, only about 100 meters, and you think you're still one km far away. Your doctor orders a 5 mg tablet, but you mistakenly take 500 mg. All these false measurements can kill you."

"What makes the correct measurement?"

"First, we need a standard measurement system that everyone agrees."

"Then?"

"We need correct rulers."

"What makes a correct ruler? In other words, if one of your rulers becomes faulty, how would you know?"

"I wouldn't know. As a user, I rely on scientists to give me correct rulers. I think you have to ask scientists that question: How can you ensure rulers always give correct measurements? If experts cannot do that, I don't know who else can. Imagine that I'm a pilot in charge of a passenger airplane one day, flying in extremely poor weather with almost no visibility. In the cockpit, in front of me, are hundreds of meters giving all kinds of measurements: altimeter, airspeed indicator, vertical speed indicator, magnetic compass, heading indicator, HSI, ADI, turn indicator, clocks, and so on. All of them must be absolutely correct. All my crew and passengers' lives depend on that. If those meters are faulty, you know the consequences."

"How much do you normally weigh?"

"Why?"

"Just give me anything."

"100 kg."

"If you jump on a faulty scale that reads 150 kg, would you know it's faulty?"

"Of course, if the reading is way off, I would know immediately. But if it's only a small margin, say 99 kg or 101 kg, I wouldn't."

"Let's say you're normally 100 kg. One day, you check on a friend's scale, which says 102 kg. How would you know whether it's a faulty scale or if you've put on weight?"

"I'd check on other scales."

"Let's say you'll check on 10 other hospital scales (which are regularly checked and tested), and they all confirm 100 kg?"

"Then I would say my friend's scale was faulty."

"Right. So even though you're not an expert, as a user, you can roughly tell which ruler is right or wrong by comparing it with other correct rulers, right?"

"Yes."

"So, talking about measurement in science, there are rules we must follow:

1.  A measurement system is defined and agreed upon by everyone.
2.  Correct rulers are used for correct measurement.
3.  All correct measurements must be the same.

Agree?"

"Yes, I agree."

"That means one object cannot have two different measurements. Agree?"

"Yes."

"No length is one km and three km at the same time. No weight is 100 kg and 300 kg at the same time. No temperature is 10°C and 30°C at the same time. Agree?"

"Yes."

"And no time-lapse is one hour and three hours at the same time, right?"

"Why?"

"It's in the definition. Measurement is to assign numerical values to physical objects accordingly so that everyone agrees. All numbers are different. We need to assign different numbers to different things. The same things must have the same numbers. Different things must have different numbers. If the same things have different numbers and different things have the same numbers, that is not measurement, and we would end up disagreeing again. Are you with me?"

"Right."

"Once we have an absolute measurement system, then we can talk about change and difference. Agree?"

"I hear what you said. But time is different."

"Look. If you work in a factory for three hours, would you like to get paid for three hours?"

"Yes."

"If you work for three hours, and the boss pays you only one hour, would you call that cheating?"

"Yes."

"Therefore, if you work for three hours according to your clock, you must want all factory clocks and your boss' clock to record the same three hours, right?"

"Yes."

"If any clock displays a different time-lapse to other correct standard clocks, it must be faulty. Agree?"

"Yes."

"And that means there's no time dilation. There are only correct clocks and faulty clocks. If time dilation exists, in the sense of clock slowing, it's a sign of faulty clocks. Clocks are meant to tell the correct time. If they fail, they're no longer clocks, but faulty clocks. Faulty clocks are to be adjusted, fixed, or thrown away. Are you with me?"

"Yes. But Einstein said time dilation exists."

"Einstein was wrong. He didn't know what time meant. Time cannot be dilated. If zero division is illegal, time dilation is illegal, too."

"I know your time is not dilated. But what if Einstein meant time to be something else? What if Einstein's time is different from Newton's time?"

"Wow, wow... what is it? Einstein's time? Is that a new term? What's its definition?"

"I don't know. But what's your definition of time?"

"Time is a standard speed that measures the speed of change. To measure, we need tools – rulers for distance, scales for weight, thermometers for temperature, gravimeters for gravity, clocks for time, and so on. Therefore, all Newton clocks must always show the same time no matter what. That's the whole idea. It's the whole purpose of measurement. It's Newton's definition of clocks, or else they're faulty. Once we have absolute clocks displaying absolute time, then we can talk about 'speed' or the rate of change. Measurements make no sense if we don't have absolute clocks displaying absolute time. How can we make a difference in life if the same things are measured as different without us doing anything?"

"Many people might disagree with your definition of time."

"Fine. What is their definition?"

"I don't know. But how do you know if your definition is right or wrong?"

"There's no right or wrong definition. Only a definition that makes sense or not, something you understand or not, and you agree or not. I call a cat a cat, but if you like, you can call a cat a dog, and it's still okay as long as we understand and agree. But make sure you remember that, so next time, you cannot switch and say a cat is a cat again, because that would confuse everyone. It's the basic principle of using language in communication.

"My definition of time is simple. Time is what correct clocks tell. Therefore, all clocks, by definition, must always show the same readings no matter what, unless they're faulty. Now, if you have a different or better definition, please tell me."

"I don't."

"What about Einstein's time that you've just said? What's his definition?"

"I don't know."

"Did Einstein ever define what Einstein's time was?"

"No."

"He must've used Newton's time – the same as everyone else then?"

"Possibly."

"Do you see it as a problem?"

"I guess so."

"Of course, it is a problem. Either he uses Newton's time and follows my rule, or he uses Einstein's time – tell me – and I will follow his rule. But first, he must clearly define *his* time, so I know what the hell he's talking about. Similarly, regarding temperature, we have the Celsius, Kelvin, or Fahrenheit system. Zero degrees Celsius is 32°F, and 30°C is 86°F. You can create any measurement system you want, as long as you clearly define it first. So again, what is Einstein's definition of time?"

"I don't know. I guess he must've meant time as something else."

"What is something else?"

"I don't know. I can only guess. There must be something else in Einstein's mind."

"What is something else that cannot be expressed in Newton's world? I have space, time, speed, and 1.7 trillion words. What more do you need?"

"I don't know."

"Let me tell you what I know. The truth is that Einstein had used my time and forgotten its rules. The whole world has used my time, and since Newton's time has been so good and efficient for so long, people forget where it came from and its associated rules. They take it for granted. They think time is a physical ticking system that already existed before humans. They don't know that time is an imaginary product we've created to measure speed. Everything has a speed. If we want to make a difference in life, change its speed. If you want to change something, change its speed. Once the difference is made, you might ask if we want to measure it. If not, it's fine; leave it. If yes, then we cannot measure speed without accurate clocks to tell time. In fact, nothing can be measured without accurate rulers. The light speed, age of the universe, frequency of color, rate of your pulse, Dow Jones Index, economic growth, population growth, and so on, for whatever change in this world, if you want to measure it, you must use the same unit of measurement. Otherwise, all measurements make no sense.

"How many times do I need to say this? Kilometer, Gb, Mhz, $m^3$, $cm^2$, kg, decibel, psi, mg, hour, and second are all measurement units. They're all manmade concepts. They're all standards we created to compare things. In Newton's world, they need to be absolute so that all relative change can be measured. Even a day might become longer in the future due to Earth's self-rotating speed slowing down. A day in the next million years could last 28 hours instead. That statement only makes sense if we keep the same value of one hour.

"Not only does everything change, everything changes at different rates. That's why we have 'speed' to measure change as 'difference over time.' And that's exactly what goes wrong with Special Relativity. Einstein was confused between speed and time. When a faulty clock slows down, its physical speed has slowed down, but not Newton's imaginary time. It's a terrible misuse of language.

"As we previously discussed, to measure correctly, we need rulers. But beware: rulers, as physical objects, regardless of how well they work now, might change tomorrow, deform, and no longer work as designed. They can break down, will break down, and give faulty readings. Scientists must know this to prevent faulty measurements, faulty readings, and fatal consequences.

"A metal ruler under extreme heat might melt and become longer. A sensor on an airplane under extreme conditions might be damaged and give a false reading. A sand clock at different locations with different gravity might slow down, resulting in different time readings. A candle clock with less oxygen might burn slower and give a different reading. An atomic clock at different gravity might have different speeds, resulting in different readings. All clocks, like any manmade machines or any physical devices with different designs, can have their original speed changed due to external changes and become faulty clocks, giving faulty readings. That can happen to anything physical in the universe. Change causes change. That's Newton's first law of motion. Science is meant to deal with those changes and discover equations to calculate those changes. But they must never alter time. Any scientific theory that attempts to change time is wrong, illegal, and nonsense. Science, including physics, must only deal with change in terms of speed. Nothing else."

"It's interesting you said that time was a measurement unit and cannot be altered. But you just said a day might become longer in the future, so does that mean you agree that time can be altered, or in other words, is relative?"

"You've distracted me with that language confusion you can easily answer."

"No. I don't know the answer."

"What do you define a day? Twenty-four hours, or a complete Earth's rotation?"

"Both. They're the same, aren't they?"

"Not always. They are the same now, but that could change in the future. The relationship between two things can change. We define one thing based on other things.

When one of them changes, the subject and its definition no longer match. That will cause confusion requiring clarification. Let's say in a million years, when Earth's rotation speed becomes slower than its current rate, say 28 hours, you can say that a day has become longer. The relationship between a day and an hour has changed. However, the length of an hour, once chosen as a measurement unit, must not change; otherwise, time measurement makes no sense. The same thing could happen with the Big Bang theory, claiming the universe started 14 billion years ago. At the start, we didn't know what the speed of Earth rotating around the Sun was. Earth could've been bigger with a larger mass then. It could've moved faster then. Earth could've taken less than a year to complete its orbit around the Sun, say 0.6 or 0.8 of the current year. Despite that, we still use the same year length to measure everything about universe history. Otherwise, what does '14 billion years ago' mean when Earth wasn't there yet and a year's length was yet defined? Can you see that?"

"But a general reader might call that change as time dilation."

"Then it's a language problem, not science. We're using the same word for different things."

"What do you mean?"

"Can you eat the rainbow?" Suddenly, Newton asked.

"I beg your pardon?"

"Can you eat the rainbow?" Newton insisted.

"No." I frowned.

"Have a look at the magazine on the floor behind you."

I turned around and saw a magazine with its front cover headline "Eat the rainbow." I picked it up, looked closely, and got the message. The front cover showed a basket of brightly-colored fruits: apples, grapes, bananas, mangos, strawberries, etc. From a distance, it looked like a rainbow. It was an ingenious ad design for a healthy lifestyle magazine.

"What does it mean?" asked Newton.

"Eat healthy. Eat fresh."

"Why did you say 'no' before?"

"We were talking about a scientific subject, and suddenly you asked me about eating the rainbow. I thought you really meant the rainbow, so I had to say no."

"So you see how scientific language is different from general language. In scientific language, a rainbow is a rainbow, but in general language, a rainbow could be fruits or anything you can imagine, as long as you understand."

"I see."

"And that's exactly the problem with Special Relativity when Einstein was confused between general language and scientific language, between relative language and absolute language."

"How?"

"Our general language is free, rich, and relative. One word can have many different and even opposite meanings. A meaning can be expressed in many different ways, phrases, and even contradictory words. When we say 'Do you drink?' quite often, we mean 'Do you drink alcohol?' When we say 'That's cool,' we don't always mean 'It's low in temperature.' When we say 'That's so sick,' we could mean 'That's great.' When we say 'It's beautiful,' we could mean 'It's so horrible. It's exactly the monster face I want for my Halloween night.' When we say 'It's true,' we don't always mean 'It will truly happen,' but we might imply 'It's true according to this calculation, but it could be wrong according to another calculation.' Similarly, when we say 'time dilation,' in our general language, it could mean many things. It could mean 'I feel like time flies,' or 'I couldn't believe that one year has passed so quickly,' 'after a facelift, she looks so much younger,' 'after my vacation, I feel that I'm 10 years younger,' 'my body clock is slow, it needs to be fixed,' and so on. Now if you want to call that relativity, then certainly you can…"

"But what's wrong with that language relativity?"

"Nothing, as long as everyone understands. When nothing is serious, nothing matters, no one bothers, then it's okay. But be careful when it's not okay."

"When?"

"When we face life-and-death situations where confusion can be fatal. That's when you need science to fix that confusion. And before science can fix it, make sure you replace general language with scientific language with clear definitions. You can say 'Eat the rainbow' on a lifestyle magazine's front cover, and no one will complain, but if you sign a medico-legal contract containing the phrase 'Eat the rainbow,' would you care to ask why?"

"Yes."

"In general language, everyone talks about time dilation, such as time flies, I want to be 10 years younger, but if you sign a legal contract paying $50,000 for a time dilation service, like the story of 'Magic Salon,' would you care to ask for its definition?"

"Yes."

"In science fiction, we all dream about time dilation, time machines, and time travel, but if your son sits for a NASA space trip to the end of the universe and returns in one week – Einstein's time, would you care to ask what the hell that Einstein's one week means in Newton's time?"

"Yes."

"And did Einstein answer that?"

"No."

"Great, now we're getting somewhere – we're back to where we were. And why do you think that is?"

"I don't know."


"Let me tell you, son. Because first, Einstein didn't have a clue what his time was; he used my Newton's time and forgot its rules. Second, he confused general language with scientific language. In general language, time dilation is a norm – generally understood as younger looks, healthy body, and extended longevity. But translated into scientific language, time dilation – as time being dilated – is wrong, illegal, and senseless. Time, as a measurement unit, cannot be dilated. It is so because we make it that way. If Einstein was serious about his theory – a real scientific theory that can make a real difference in life, he had to state at the very beginning that time dilation doesn't exist. Time dilation is a misconception.

"'Eat the rainbow' exists in general language, but in scientific language, it doesn't. 'Time dilation' exists in general language, but in scientific language, it doesn't. You cannot eat the rainbow, and you cannot alter time. This is science. Use the correct language."

"But what if Einstein meant time dilation to be younger looks, healthy body, and extended longevity?"

"Then by all means, say that, please. Einstein could've said exactly those words – younger looks, healthy body, and extended longevity as the results of Special Relativity. And did he say those words?"

"No, he didn't."

"Why?"

"I don't know. Maybe he didn't mean them."

"Look. If one day you walk out on the street and someone who mistakes you for some celebrity comes asking for your autograph, if you're an honest man, why don't you say to that fan, 'Sorry, but I'm not that celebrity?'"

"I don't know."

"If one day you discover a quick way to make lots of money and fans come over after your discovery, and if you're an honest man, why don't you tell them, 'Sorry, but the money I've just photocopied and printed could be illegal?'"

"I don't know."

"Einstein wrote Special Relativity. If Special Relativity is not about delivering younger looks, a healthy body, and extended longevity as expected and admired by the general public, why didn't Einstein say, 'Sorry, but my time dilation has nothing to do with those?'"

"Maybe Einstein didn't even know."

"Why didn't he?"

"Because as a scientist, he knew his equation has no variables representing longevity."

"Exactly. Then what does his equation represent?"

"Time dilation."

"What did he mean by time dilation?"

"Maybe he meant clock slowing?"

"Which clock?"

"He never said which one."


"You know what? As soon as you mention 'clock,' you open up more problems than solutions, more questions than answers, and more confusion than certainty. What kind of clocks do you prefer? Rolex, Omega, Casio, Sony, Apple, analog, digital, sundial, pendulum or sand clocks? Are clocks all the same? Are clocks all manmade? If so, what kind of clocks do you want to make? What speed do you want clocks to have? Do you want to create correct clocks that never run slow, or faulty clocks that will run slow? If you know exactly what makes clocks faulty, why don't you use that knowledge to stop clocks from becoming faulty? Or do you still want to create faulty clocks just to prove that you're right that clocks can be faulty? If you want to make spaceships, but your spaceships cannot fly, and you know exactly why they can't fly, why don't you use that knowledge to make them fly? Do you really know what you want to do, or don't you? Does Einstein know what clocks are, or doesn't he? Do any scientists want to use faulty devices in measurement, or don't they? And what about non-clocks? Why does time dilation apply only to clocks but not non-clocks? And if time dilation can apply to all clocks and non-clocks, does 'time dilation' mean 'speed change'? Yes or no? If the answer is yes, does that mean inertial motion can cause speed changes in objects' behavior against the first postulate? If the answer is no, does that mean time dilation can never be realized, and all evidence allegedly confirming time dilation is just garbage? Do any scientists know what time dilation is? Does Einstein know? Does anyone know?"

I became confused. I was lost. It was not a good sign. "I wish he was here, so I could ask."

Newton looked at me, saying, "Don't wish. He wouldn't know. No one in your world knows. Come back next week, and I'll teach you a lesson about clocks."

# CHAPTER 7
# ALIEN FROM X

*If I experiment enough, I get a deeper understanding.*
*TERENCE TAO*
*Mathematician, child prodigy*

Newton didn't talk about clocks the following week. "Instead, I want to tell you three stories first," said Newton.

"Not again. Why?" I protested.

"It's too early to talk about clocks. Since we last met, I've realized some basic things you need to understand first; otherwise, we keep falling into the same sinkhole."

"What are they?"

"Three stories: 'Alien from X,' 'Moon Clock,' and 'Illusion of measurement.'"

"I'm listening."

"Imagine that one day, an alien from Planet X crash-landed on Earth," Newton started with the first story.

"Where is Planet X?"

"Stop asking wrong questions. Focus on what matters. What doesn't matter doesn't need your focus."

"Okay." I gave in again.

"The alien looks exactly like a normal human, with eyes, ears, and hands, like all of us. Except that he's huge, 20 meters tall, weighing one ton, and extremely slow. Everything about him is slow. His speech, movement, eating, breathing, blinking, sleeping, and so on. We try to

teach him English, and he can learn, but the problem is his slowness. It takes him one minute to say, 'How are you?' at a very low pitch. And when we talk to him, he cannot pick up any sound because, according to him, we speak too fast. To communicate, we have to talk using a speech delay device, which makes our speech 50 times slower. He complains that everything around him moves so fast it hurts his eyes, even when we just walk at a normal pace. If we run, he sees nothing. It takes him 20 seconds just to blink. It takes him four hours to finish a meal. He sleeps non-stop for 30 hours. His pulse is about 20 beats per minute. His typing speed is about two words per minute. His skin is very thick. If you run boiling water over his hand, he feels nothing. Put fire on his arm; he doesn't feel hot until two minutes later."
(See Figure 7.1.)

*Figure 7.1 Alien from X*

"In short, he's a perfect example of 'time dilation'?"

"Yeah, I guess you could say that. But there are exceptions. When he falls, he still falls at the same speed as everything else (given the same air resistance) on Earth. He doesn't fall slower despite his slower body clock. When he talks, words coming out from his mouth are very slow, but once they're out, the sound still travels at the same speed as other sounds on Earth. Are you with me?"

"Of course. Why do the laws of physics have to be different just because his body clock is slower?"

"You're absolutely right. And another thing, since his arrival on Earth, he has become of huge interest to all scientists, media, and the public. Everyone wants a piece of him. Everyone wants to see him, touch him, or interview him. To help him keep up with so many appointments, we give him a watch. What sort of watch do you think we would give him?"

"The same watch as everyone else's."

"Exactly. He needs to tell time like everyone else. Therefore, regardless of how slow his body is, he needs the same clock as everyone else."

"Can he read the minute hand?"

"He struggles even with minutes. So he mainly relies on the hour hand. That's why when you ask him what time it is, even if he can give you the correct time by looking at his watch, it's already a few minutes late by the time he finishes saying it."

"Why are you telling me this story?"

"There is no time dilation. Time dilation is just a word for a headline, a sensation, and attention-grabbing. But semantically, it's speed change. Nothing else. The story about the alien from X demonstrates a situation where some speeds slow down relative to something else. His breathing is slowed down. His blood circulation is slowed down. His speech, pulse, feeling, aging, hair growth, and so on are all slowed down. But they aren't slowed down at the same rate. And some might not be slowed down at all, such as the speed of gravity, or the speed of his body falling onto the bed, the speed of sound, the light speed, the speed of a mosquito flying around his head, the speed of water streaming down his body, and most importantly, the speed of clocks, or the speed of the watch he's wearing. Everything can slow down or speed up. Speed can change; time cannot. Einstein was confused about time and speed. If we want to measure speed and make sense of our measurement, keep the same time. Time dilation is a misconception."

I didn't say anything for a while. "You know, your story is like a thought experiment."

"It is."

"Thought experiments prove nothing because they're just thoughts. They're not real."

"You're right. I'm glad you said that. Thought experiments are like a scripted movie. Events in a thought experiment happen the way we want them to happen. All scripts are right because actors are paid to make them right. Therefore, a thought experiment cannot prove reality; it can only explain your dream of reality. It tells a story. The story I tell has the title 'Speed.' Someone else tells exactly the same story titled 'Time dilation.' Same story, two different titles. Despite its sensational sound, time dilation is misleading. That makes Special Relativity a theory of misconception."

# CHAPTER 8
# MOON CLOCK

*Never mistake motion for action.*
*ERNEST HEMINGWAY*

We met again next week for the second story.

"Imagine far away in the galaxy lies an Earth-like planet with similar humans," Newton started. "There are some differences, though. All planets in that solar system still revolve around their sun but never rotate themselves. Therefore, one side always faces the sun, and the other always faces darkness. Humans there only live on the dark side since the other half is extremely hot. That means that all their lives, humans there live at night, under a starry sky. Apart from that, their life is similar. They have electricity, cars, buildings, like everything on Earth."

"Do they have moons?"

"No… Well, actually, they do. It's another twin planet that always floats around right in the sky like a huge moon. It's so big and so close, 1000 times bigger than our moon. That means whenever humans there look up outside, they'd always see a starry night sky with a huge full moon."

"What else?"

"People there are happy, and they love traveling. The planet has thousands of islands, amongst which is one called 'Nevertime' island."

"Why?"

"A scientist discovered that empty island in the last century. He built a lab to conduct some experiments. After an explosion accident, the island changed. 'Time' slows down there now, about two or three times slower than other islands. Any humans or animals there move slower, age slower, eat slower, and breathe slower."

"Like Einstein's time dilation?"

"Yes. But not because of motion. Motion is not action."

"How?"

"They still don't know. Different air, different radiation, different electromagnetic fields, or whatever. What they know is very soon after they get onto the island, their bodies start slowing down. You live there for one year of normal time, come back to your world, and realize two years have passed. One artist who often finishes 10 paintings a year, staying there for two years, comes back with only 10 paintings. One mathematician, who often produces an equation each year on the mainland, leaves his town for two years to live on the island and returns with only an equation completed. You normally can run at 10 km/h, but now you can manage only five km/h on the island."

"But do you look younger upon return or not?"

"Yes."

"So it's real 'time dilation'?"

"Yeah, you can say that."

"What about the twin paradox?"

"Yes, it happens there too. There are two identical twins, one of whom is sick of being mistaken for his brother. So, one day, he decides to leave his brother for ten years to live on that 'Nevertime' island. When he comes back, he has become different. He definitely looks five years younger than his remaining brother, even without a new birth certificate."

"What about his height, his waist, or his shape?"

"No comment. It's not a topic here."

"You said before that all humans and animals would slow down, but what about non-living things? Would they slow down as well?"

"Oh no. They don't change at all. Similar to the story of 'Alien from X' – gravity, light, and sound speeds are still the same. When they shower, water still runs down on their body at the normal speed."

"But since their body slows down, they must notice water streaming down faster?"

"Exactly."

"Apart from that, while living there, they cannot tell if time slows down on that island, right?"

"They can."

"How?"

"I forgot to tell you one thing. No one there uses watches."

"No watches? Do they work? Do they need to be on time? Do they need to keep appointments?"

"They do."

"How can they tell time without watches?

*Figure 8.1 Moon Clock*

"Just look up. The full moon is always there. So, to keep time for everyone, they built a super gigantic clock on the moon so everyone could see – the only clock on the whole planet. Wherever you go, which island you're from, even when you're in 'Nevertime' where everything slows down, if you want to check the time, the real time that everyone uses, so that you can tell if your body slows down or speeds up, just look up in the sky and you see. You know what it's called?"

"Moon clock."

"You got it." (See Figure 8.1.)

"Why are you telling me this story?"

"It looks like time dilation, but again, there is no time dilation. Time is only an imaginary concept – a standard absolute speed – humans have created to measure other speeds. Without absolute time, the measurement of speed makes no sense. What really happens on that 'Nevertime' island is not time dilation, but speed dilation or speed change. 'Speed change' is the correct scientific term to describe the difference between that island and others. Nothing else. Everything has a speed. Some change, some don't. And they don't change at the same rate.

"In life, if a scientist can make a difference by changing some speed for a better cause, it's great. Whether that be the speed of hair loss, the speed of bone density, the speed of facial wrinkles, the speed of thinking, the speed of muscle waste, the speed of reflex, the speed of aging, the speed of death, the speed of sound, the speed of gravity, the speed of freefall, the speed of light, the speed of growth, the speed of wealth, the speed of the economy, the speed of a tape rewinding, the speed of recovery, the speed of understanding, the speed of awakening, and so on.

"But regardless of how our speeds change, there has always been, and will always be, the need to have accurate clocks displaying the same time. Call it Newton's time, absolute time, proper time, universal time, standard time, imaginative time, perfect time, or whatever name you want – still, it is the purpose of measurement. How we can find a perfect representation of time is another question. But that's the idea.

"Einstein was confused between time and speed. Speed can change, but not time. By changing time – the unit of standard speed – Einstein has illegally altered the measurement system. Scientifically, it's wrong. Ethically, it's misleading. Socially, it's cheating. It has fooled lots of people, including many scientists, that inertial motion can change the world. It's a total contradiction of Special Relativity's first postulate, which states that inertial motion alone cannot change the world. Special Relativity is a theory of self-contradiction."

# CHAPTER 9
# ILLUSION OF MEASUREMENT

*People don't want to hear the truth,*
*because they don't want their illusions destroyed.*
*FRIEDRICH NIETZSCHE*

"Can we change the world by changing its measurement units?" asked Newton.

"No," I said.

"Imagine we have a new measurement system – a shorter ruler that measures a one-meter length before as two meters now. What's the new measurement of a standard two-meter-high door?"

"Four meters."

"Does the door change at all?"

"No."

"But if you look at the reading, from two meters to four meters now, that surely gives an impression of a higher door, right?"

"Yes, it's misleading and confusing. We cannot change the world by changing its measurement system."

"How would you know if the world has changed?"

"When its measurement changes – only if it's correct measurement within the same measurement system." (See Figure 9.1.)

"You play tennis?"
"Yeah, but I'm no good."
"How fast can you serve?"

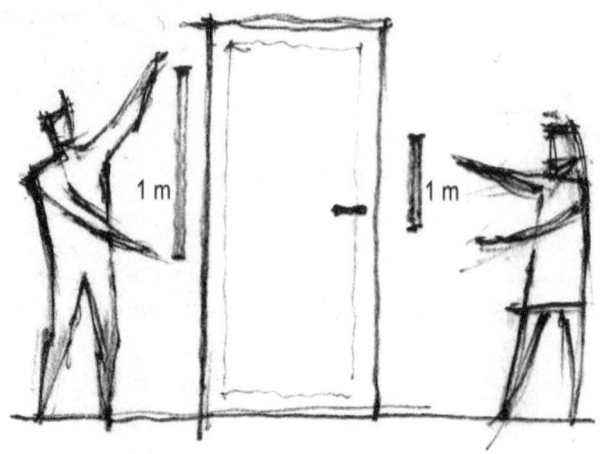

*Figure 9.1 Same door, different measurements*

"One hundred km/h."

"Can you ever serve at 200 km/h?"

"Not a chance."

"What if you could?"

"That would be a miracle. I'd beat anyone in my tennis club if I constantly had that serve."

"Is there another way?"

"Not a chance. Unless we use that shorter ruler – that measures a 100 km length as 200 km now."

"But that's cheating."

"It is."

"Talking about cheating, what if, instead of a shorter ruler, you use a slower clock which measures one second before as half a second now?"

"Same result. That means I can serve at 200 km/h."

"What if you use both a slower clock and a shorter ruler?"

"Shorter ruler or longer ruler?" I asked.

"Yeah, whatever. Say, it's a longer ruler that is twice as long."

"That would offset any change. My serve's measurement would go back to 100 km/h."

"Let's just make one change at a time – use that slower clock only. You normally drive an hour to work. How long is the drive now according to that new clock?"

"Half an hour."

"You normally work eight hours a day. How long would you work now?"

"Four hours."

"You normally work 40 hours a week. How long would you work in a week now?"

"Twenty hours."

"Let's say you get paid $100 per hour. How much is your pay rate now?"

"Two hundred dollars per hour."

"You normally run at three km/h. How fast can you run now?"

"Six km/h."

"A flight from New York to Sydney normally takes 20 hours. How long does it now?"

"Ten hours."

"Wow, it sounds fantastic, doesn't it? You serve better. You run faster. You work less hours. You earn more. You travel in half time. All of a sudden, the world has become better."

"Yeah, but it's cheating. Same world, different reading."

"You said 'different reading.' Does everything now have a different reading?"

"Yes, I guess so."

"How many minutes are there in an hour, before and after?"

"Still 60 minutes."

"So, that doesn't change?"

"No. I was wrong when I said everything would change."

"How many seconds are there in a minute now?"

"Still 60 seconds."

"You normally go to work at 9.00 am. What time would you go to work now?"

"4.30 am."

"So, that's changed, right?"

"Yes."

"What does it tell you?"

"Not everything changes because of one change. Some change, some don't."

"Why?"

"I think it's a relationship thing. Within a group of related things, one change would impact others. But if they're not related, nothing would change."

"I see. You normally break for lunch at 1.00 pm. What time would you have lunch now on that slower clock?"

"6.30 am."

"You normally finish work at 5.00 pm. What now?"

"8.30 am."

"You normally have dinner at 7.00 pm. What now?"

"9.30 am."

"You normally go to bed at midnight. What now?"

"12 o'clock."

"12 pm or 12.00 am?"

"12 o'clock. No need to say am or pm because a day has only 12 hours now."

"What happens in the next hour? Is that a new day or still the same day?"

"New day."

"Right, so a 24-hour day now only has 12 hours on a slower clock. What about months? Say, a month normally has 30 days; what now?"

"Still 30 days."

"How many days are in a year now?"

"Still 365 days, except leap years."

"So how older would you be after a year on both clocks, old and new?"

"Still one year older."

"So you still age the same despite a slower clock?"

"Yes."

"So where is time dilation?"

He caught me by surprise. "I don't know. But why do they say if your clock runs slower, you would age slower?"

"Who says that?"

"Einstein and almost all mainstream physicists who support Special Relativity and the twin paradox."

"Remind me again about the twin paradox."

"A traveling twin, after a high-speed trip, would come back and see himself age slower than his twin brother, who remains on Earth. At a certain high speed, his clock would run 70% slower, meaning if he's away for 10 years according to an Earth clock, he would age only seven years."

"But during those 10 years, how many times has Earth orbited around the Sun?"

"Ten times."

"If Earth sends an email to him every Christmas during the trip, how many emails would he receive?"

"Ten emails."

"So that means they've all aged the same 10 years, haven't they?"

"Yes, I would think so."

"Where does that twin paradox's prediction come from?"

"Einstein's time dilation equation: $t = t'\sqrt{(1 - v^2/c^2)}$."

"What are t and t'?"

"Let's say t as time registered on the traveling clock and t' as time registered on Earth clock."

"Right, so it's time registered on a clock. Does t relate directly to the speed of the traveling clock, and t' the speed of Earth clock?"

"I would think so."

"Are they both for the same time-lapse starting when twins separate and reunite later?"

"Yes."

"t is not equal to t', so that means two clocks have different speeds for the same time-lapse?"

"Yes."

"Is it legal? Can you use two different rulers to measure the same thing?"

"No."

"Can you use two different clocks with different speeds in one experiment to give an impression that you've made a difference?"

"No."

"If you use a slower clock, surely you would obtain a different reading. But would that change the world?"

"No."

"If your clock runs 70% slower, would that make Earth orbit the sun 70% slower?"

"No."

"Would using a slower clock make Earth rotate slower?"

"No."

"Would it make you age slower?"

"No."

"It might give you all different readings about speed. But would it give you a different age?"

"No."

"Again, where is time dilation?"

"I don't know."


"Let's say it would take 1000 hours to build a house. How long does it take according to that half-speed-slower clock now?"

"500 hours."

"Does it actually take any less time?"

"No. It's just a different reading. Nothing physically changes."

"Let's say it would take 1000 hours or approximately that. Six months to build a house (based on an eight-hour work per day, 40-hour week). How many months does it take if you use that slower clock now?"

"Still six months, according to my calendar."

"No. I mean, according to the built-in calendar inside your slower clock."

"Oh, it'd be different then. My clock's built-in calendar wouldn't know a day has only 12 hours now. So after two days on normal clocks, my abnormal clock still reads one day."

"So what happens after six months on normal clocks?"

"Then it's three months, according to my abnormal clock's calendar."

"Why did you say six months before and now three months?"

"I was confused before. Now I know six months in your paper calendar is three months in my clock's calendar. Same time, different reading."

"Does that mean time reading is dependent on clocks and calendars?"

"Of course. What does time mean without clocks and calendars? Can we say let's meet at 3.00 pm the same day next month without reference to any clocks or calendars?"

"So does that mean if we use absolute clocks, we'll have absolute time, relative clocks give relative time, different clocks give different times, correct clocks give correct time, and faulty clocks give faulty time?"

"Yes, it's that by definition."

"What if you use a frozen clock? According to that clock, how long does it take to build that house?"

"Zero days."

"That means you can build that house *instantly*?"

"Yes, according to that frozen clock."

"Does that mean regardless of three months, six months, or three years, that clock still registers any time-lapse as zero, meaning instantly?"

"Yes, according to that faulty clock."

"Faulty? Why do you say it's a faulty clock?" asked Newton.

Again, it surprised me. Newton pressed on. "Why do you say the frozen clock is faulty, but not just another clock with a different speed?"

"Because three months, six months, or three years are different to me. If a measurement device fails to detect that difference, which matters to me, I'd say it's faulty. Whereas, other clocks, despite their different speeds, still show me some differences to which I can relate."

"What's your definition of faulty measurement?"

"Measurement that doesn't match reality. It gives me faulty expectations. In extreme cases, it could kill me. Say, I'm a builder, and time is important to me. Building a house in three months, six months, or three years makes a huge difference. I might have some holiday bookings, family commitments, personal business, and I need money. Time matters to me. I don't care what clocks or what type of calendar aliens use. I don't care if my three months mean 30 months to someone else flying in the sky using a different calendar, but I expect everyone with whom I interact to use the same device and language I use. Three months means three months, and I want full pay for my three-month work according to the legal contract."

"It would be a totally different story if you could finish the house in real six weeks instead of six months, say, with super-fast robots?"

"Of course. Life is wonderful if no one cheats – no one uses different clocks, different calendars, different rulers, or different meanings despite the same words."

"Similarly, it would be wonderful if you could improve your serve from 100 km/h to 200 km/h based on the same measurement system, right?"

"Yes, of course."

"Why would anyone want to use a different kind of measurement?"

"To create an illusion."

# CHAPTER 10
# WHEN SCIENTISTS MURDER

*There is not a single concept of which I am convinced that it will survive, and I am unsure whether I am on the right way at all.*
*ALBERT EINSTEIN*

"Can illusions kill?" asked Newton.

"Yes," I said.

I'd thought he would talk about clocks today, but again he didn't.

He paused for a few seconds, then changed direction. "What do you call a liar who doesn't lie?"

"A standing liar?"

"It's not a joke."

"Why would you call someone a liar if she or he doesn't lie?"

"I'm asking you."

"You mean a liar who lies to you, but never lies to everyone else?"

"No, he never lies to anyone."

"I think it's a self-contradictory statement. It's nonsense."

"What would you call a cancer patient who doesn't have cancer?"

"Nonsense."

"What would you call number five that is number seven?"

"Nonsense."

"What do you call a triangle that has four sides?"

"Nonsense."

"What do you call a convicted murderer who never murdered and never got convicted?"

"Nonsense."

"What would you call a man who is a woman?"

"Can a man be a woman?"

"I'm asking you."

"I think it's all about definition. Absolute definition gives absolute meaning. Relative definition gives relative meaning. Loose definition gives loose meaning. Contradicting definition gives contradicting meanings. So, depending on your definition of man or woman, the answer could be yes or no or nonsense."

"I see."

"But why do you ask me all this?"

"You need that to understand my next story."


Newton suddenly changed his tone.

"The year is 3001. A tragic accident happens when a river bridge completely collapses under heavy flood, and 17 children drown. The tragedy could have been avoided if a 20-meter-long floating tube had been dropped into the water in time to block the river, which is also 20 meters wide, so the children could cross over. Many spaceships flying over the river at that time received calls for help and dropped down floating tubes. But they're all the wrong size – 15, 16, or 17 meters. They all tip over at a huge waterfall downstream and break apart. Only one was the right size of 20 meters, but it was delivered too late, and some children had already drowned. An inquiry is conducted to find the cause of the tragedy. All scientists working in those spaceships were called to give evidence. Questions asked, answers given, and disturbing stories heard…"

I said nothing. He continued.

"Did those spaceships know they needed to drop floating tubes long enough to block the river?" (See Figure 10.1.)

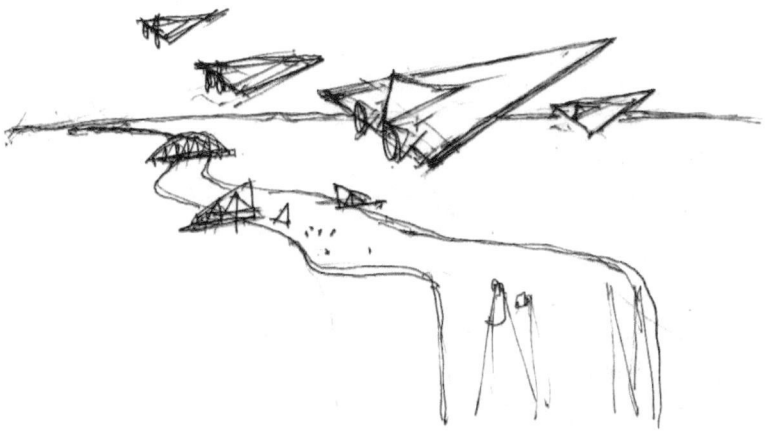

*Figure 10.1 Can scientists murder?*

"Yes."

"Had those tubes been dropped in time, would they have saved the children?"

"Yes, there was a 90% chance all the children would've survived."

"Why didn't that happen?"

"Those floating tubes they dropped were too short."

"Why? Can't they measure the river's width from their flying spaceships?"

"Yes, they can."

"What was the measurement they obtained?"

"15, 16, and 17 meters."

"But the river is 20 meters wide at that section."

"Yes. However, according to Einstein's Special Relativity, that 20-meter length should be measured as 15 meters by the spaceship traveling at 200,000 km/s. Twenty meters in this reference frame is 15 meters in another inertial reference frame."

"So it's a faulty measurement, isn't it?"

"Well, if you look at it that way…"

"What other ways could you possibly look at? Children are dying. They desperately need a 20-meter-long tube. But the scientists flying up there miscalculated it as only 15 meters."

"Actually, they know the river is 20 meters wide according to the ground people."

"Beg your pardon?"

"According to Special Relativity, 15 meters is the river's width measured by the spaceship flying at 200,000 km/s (0.66 c). But knowing their flying speed, the spaceship crew can still use Special Relativity's equation to work out the river's original measurement relative to the ground, which is 20 meters."

"Why didn't they use that measurement?"

"They got confused about which measurement they should use. One spaceship's captain first ordered his staff to prepare a 15-meter tube, but as soon as the spaceship slowed down to 150,000 km/s (0.5 c), the order changed to 17 meters, then again to 18 meters."

"Are you serious?"

"Yes."

"You can't be serious. Are you talking about different spaceships?"

"No, even for the same spaceship, but at different speeds, its measurement keeps changing."

"For the same object on the ground?"

"Yes."

"Does the river's width change at all?"

"No."

"The same object doesn't change, but its measurement keeps changing. What the hell are you talking about? Don't you call that a faulty measurement?"

"Well, if you look at it that way…"

"What other ways can you possibly look at?"

"It might be confusing, but according to Einstein's Special Relativity, space is relative to speed. Length contracts when it moves."

"Does the river fly?"

"No. But the river can be seen as flying while the spaceship is stationary. It's Relativity."

"Is Relativity the reason 17 children died?"

"I guess that's one way to look at it. It's more about which measurement is used for which purpose. As I said before, technically, we can use Special Relativity's equation to calculate the river's original width as 20 meters. And apparently, one spaceship did drop a 20-meter-long tube at the last minute."

"Why did that spaceship drop that 20-meter tube?"

"Because that spaceship's captain, after some initial confusion, realized they must drop a length according to the users' needs. The length, according to ground people, is 20 meters, so they must drop a 20-meter length."

"Right, I see. So, finally, our wonderful scientists realize the measurement is for users – for real people with real purposes in real life. Who else is measurement for, if not for users? For tourists? For dreamers? For poets?"

"No. For observers."

"For observers to have fun?"

"No. But Special Relativity allows moving observers to have that kind of shorter measurement."

"What for?"

"That's how measurement is done in Special Relativity."

"But what for?"

"I don't know. It's Einstein's idea. We're just followers."

"What about stars millions of light years from Earth while Earth, the galaxy, and everything else are moving? Are their dimensions calculated the same way? Are they for observers only and not users? And when users arrive there one day, they'd realize those measurements are all fake?"

"I don't know."

"Do you think those scientists using Special Relativity's measurement should be charged with murder?"

"I don't know."

"Let me tell you that story from another perspective," Newton said. "The year is 3001, and a tragic accident happens to a high-speed spaceship when its underneath door falls off during flight, and 17 scientists are thrown outside into space to their deaths. When that belly door falls, the crew can only hold up inside for only a few minutes until a new door is delivered while the spaceship is in auto-cruise mode. Emergency calls are made to the ground, and a new door is delivered in time to the flying spaceship, but it's the wrong size – only 15 meters long, while the belly door is 20 meters long (flight direction). An inquiry was conducted to find the cause of the tragedy. All the surviving scientists working in that spaceship and ground station are subpoenaed to give evidence. Questions asked, answers given, and disturbing stories heard." (See Figure 10.2.)

Again, I said nothing. He continued with the story.

"Why didn't the spaceship stop flying after losing its door?"

"The auto-cruise system got jammed."

"Despite that, if a correct door got delivered in time, could the tragedy be avoided?"

"Yes, 90% chance they would survive."

"Why didn't that happen?"

"Wrong-sized door delivered. It doesn't fit."

"What's the door's measurement according to the spaceship crew?"

"Twenty meters long."

"What measurement did the spaceship request?"

"Twenty meters long."

"So the ground station people knew?"

"Yes."

*Figure 10.2 Can faulty measurements kill?*

"Do they have other spaceships of the same model on the ground?"

"Yes."

"What's the measurement of those doors?"

"Twenty meters."

"So the scientists at the ground station knew?"

"Yes."

"What's the replacement door that they actually delivered to the spaceship?"

"Fifteen meters long."

"Why?"

"The spaceship travels at 0.66 c. According to Special Relativity, its 20-meter-long door was contracted to 15 meters."

"If their door contracted to 15 meters, why would the spaceship crew measure their door as 20 meters and request a 20-meter door?"

"The spaceship crew would not know of their contraction, because their rulers also got contracted, according to Einstein."

"Prior to the accident, did the flying spaceship communicate with the ground about their flight?"

"Yes."

"Did the spaceship crew report their length contraction?"

"No."

"Did they feel their length contraction?"

"No."

"Did they ever take selfies before and send them to the ground?"

"Yes."

"Did it show any length contraction in those photos?"

"No."

"Did they ever film themselves and send them to the ground?"

"Yes."

"Did it show any length contraction in those video clips?"

"No."

"Why?"

"The laws of physics remain the same in inertial motion, according to Special Relativity's first postulate."

"In short, is there any evidence of length contraction following Einstein's Special Relativity?"

"No. Only Einstein's calculation."

"Is there any evidence for that calculation?"

"No."

"Are you sure?"

"Hang on, there is actually one."

"What is it?"

"There is evidence from a thought experiment. If you stand outside the spaceship looking into its window, you would see everything therein contracted."

"The spaceship is traveling at close to light speed, so how can you possibly see anything inside through its window while standing outside?"

"That observer must travel at the same speed as the spaceship then."

"So are you standing still or flying?"

"I'm not sure."

"What's different if you stand outside or inside the spaceship if you travel along with it?"

"Nothing."

"Do you have a photo of length contraction as evidence?"

"No."

"So it's all about imagination, isn't it?"

"I guess you can say that."

"What did you say before about a liar who never lies?"

"Nonsense."

"What did you say before about a cancer patient who has no cancer?"

"Nonsense."

"What did you say before about a convicted murderer who never murdered?"

"Nonsense."

"What do you say now about a length contraction that is not length contraction without any evidence of length contraction except its calculation and imagination?"

"You remind me of Schrödinger's cat."

"What the hell is that?" Newton frowned.

"In quantum physics, a cat can be alive and dead at the same time."

"What the hell does it have to do with the length contraction we're talking about here?"

"I'm sorry. I know it's way off. But I cannot help thinking about that when you mention something that is not itself. What is it that is not what it is?"

"It's nonsense, you said."

"No. Even nonsense cannot be no-nonsense."

"So what is it?"

"It's quantum physics."

"I hate this. But since you mentioned quantum physics, let me say something briefly. It's not quantum physics that you're talking about; it's nonsense. Reality is not wrong, but their description of reality is wrong. Be careful, very careful of the language some professionals use. If five is not seven, a cat can never be alive and dead at the same time. What quantum physics means is that if a cat is hidden from an observer, what happens to its fate is something the observer doesn't know. So, while he doesn't know, he can only guess that either the cat is alive *or* dead at a given time. But to say a cat that is alive *and* dead at the same time is a wrong statement in science. Many have misinterpreted quantum physics. Their explanation of quantum physics is nonsense – not quantum physics. Also, beware of relative language used by some professionals with no reference to our real world. They might use the same words, but have different meanings to ours. We use real numbers; they might use imaginary numbers. We talk about reality; they talk about illusion. We talk about life and death, which matter to everyone; they only care about right and wrong, which matter to their theory. It's like a man who smokes ice, running out to the streets, screaming that cars can't kill him. In his world, $1 = 2$, $a^2 = -1$, a cat can be alive and dead, everyone can fly, and you can live in many parallel universes. Truth is, he's absolutely right in his own world. Even when he gets run over by a truck and dies, he's still right, because now he's alive in heaven. But his truth is not our truth. Alive in heaven means dead on Earth. We speak different languages."

"But if—"

"Enough of quantum physics. Where were we before?"

"We were talking about length contraction."

"Yes. We're talking about a liar who never lies, a cancer patient who has no cancer, a convicted murderer who never murdered, a length contraction that is not length contraction, and most of all, a 'true' calculation that is so wrong, causing a fatality. What would you call that calculation?"

"I don't know."

"Why don't you call that nonsense?"

"I don't know."

"Do you think those scientists using Special Relativity's calculation should be charged with murder?"

"I think the problem is not the calculation, but the application of that calculation."

"Gosh. What's the difference between 'calculation' and 'application of calculation'?"

"Calculation is calculation. Application is application."

"You mean some calculations are meant for fun, not reality?"

"Maybe."

"Because if you apply to reality, someone would be killed?"

"If you misapply it, yes."

"How can you not misapply it?"

"Just don't apply it to reality. Einstein's length contraction is not for reality."

"What is it for, then?"

"To prove that, in theory, Einstein is correct about absolute light speed."

# CHAPTER 11
# NONSENSE OF THE SECOND POSTULATE

*Most of what I do does not save lives.*
*TERENCE TAO*

*Mathematics is a language.*
*WILLARD GIBBS*
*Yale's first professor of mathematical physics*

*There is no actual limit on particle speeds.*
*What exists instead is a visibility limit:*
*If a source and a receiver are in mutual retreat*
*at total relative speed in excess of 2c,*
*then signal contact and reception cannot be achieved.*
*CYNTHIA WHITNEY*
*PhD from MIT in Physics and Relativity, now a critic of Relativity*

*In Galileo's time it was heresy to claim*
*there was evidence that the Earth went around the Sun,*
*and in our time it is heresy to argue that there is evidence that*
*the speed of light in space is not constant for all observers.*
*BRYAN WALLACE*
*Physicist*

"Do you know what Special Relativity's second postulate is?"

"Yes, light speed is absolute regardless of the source and the observers' speed."

"That's the summary. The postulate consists of three separate statements that many people are not fully aware. Can you separate them?"

"1. Light speed is always constant, approximately 300,000 km/s (in a vacuum).

2. Light speed is always constant regardless of the source's speed.

3. Light speed is always constant regardless of the observers' speed."

"What's the difference between those three statements?"

"The first two statements are about the constant light speed relative to a single observer. The last one is about the constant light speed relative to multiple observers regardless of their motions."

"You see the difference, but do you understand the difference?"

"No."

"How did Einstein know the light speed is 300,000 km/s? Did he measure it?"

"No. But are you disputing that value?"

"I'm not disputing that value; I'm questioning if Einstein understands the meaning of that value. Did he have any devices to measure the light speed?"

"No."

"Did he ever attempt to measure the light speed?"

"No."

"Then how did he know the absolute value of light speed?"

"He just followed the result obtained by other physicists."

"I see. So if other physicists measured light speed at 200,000 km/s or 400,000 km/s, would Einstein have accepted it?"

"I guess so. As I said, it's not Einstein's role to measure light speed. That measurement was obtained by other engineers, technicians, and testers."

"Did Einstein ever use atomic clocks to test time dilation?"

"No."

"Did he ever conduct any tests to prove length contraction due to high-speed travel?"

"No."

"Did he ever conduct any tests to confirm clock slowing due to inertial motion?"

"No."

"Did he ever conduct any tests to see if light can be at two different locations at the same time?"

"No. But why should light be at two different locations at the same time?"

"Do you think it's possible?"

"No way."

"Why?"

"Because if so, we cannot say light has a speed. If something has two different speeds, it must be two things, not one thing."

"Did Einstein say light has a single speed?"

"Yes."

"How do you know he said that?"

"He always said light speed (in a vacuum) is absolute (approximately) at 300,000 km/s, not 250,000 km/s, not 420,000 km/s, and not 632,127 km/s. That means he agreed that light has only one single speed, meaning all other speeds are wrong, meaning a photon of light can only be at a single place at a single time, meaning light cannot be at two different places at the same time."

"Then how did Einstein know light speed still is the same regardless of observers' motion?"

"He just made it up."

"I see. So even though Einstein had never done any tests about light speed, he could still come up with that second postulate, and everyone accepts that truth?"

"That's why he's a genius."

"You said you play tennis?"

"Yes."

"And you said you can serve at around 100 km/h?"

"I guess so. I've never actually measured it."

"Let's take it as 100 km/h. Would it be 100 km/h all the way from your racquet to the ground?"

"No. It may be 100 km/h leaving the racquet, but down to 95 km/h passing the net, then 85 km/h landing on the court, and 40 km/h after bounce."

"So the speed of the ball changes during its course?"

"Of course, air friction."

"Can you measure the ball's speed after it has landed on the court, and say it's the same speed as when the ball first comes off the racquet?"

"No way."

"Then why do we measure landing light on Earth and say it's the same traveling light from a faraway galaxy?"

"But how can we measure light before it reaches us?"

"So, because we cannot measure traveling light, we have to measure post-landing light and assume both are the same?"

"I think that's an assumption we have to make. I mean, it's easy to measure a tennis ball. We can measure the ball well before the ball reaches us by using a camera, but we cannot measure light before light reaches us. That's the problem."

"So it's a measurement failure?"

"I guess you can say that."

"Could that measurement failure lead to our faulty view of the world?"

"Maybe."

"Could the Michelson-Morley experiment be just a measurement failure?"

"Maybe."

"Can you prove Einstein's second postulate using math?" asked Newton.

"No, you can't."

"Why?"

"It's a postulate."

"What is a postulate?"

"Something you just accept as truth without proof. You can never prove a postulate. If you could, you would not call it a postulate."

"I see. But why would Einstein choose absolute light speed as a postulate?"

"He tried to explain the Michelson-Morley experiment's negative result."

"Are you sure?"

"No, I'm not. Some books say Einstein formalized that postulate even before the Michelson-Morley test and that he made it up all by himself independently."

"Whatever, he still made it up, didn't he?"

"Yes."

"So regardless of Einstein's second postulate having anything to do with the Michelson-Morley test, can we still say that the experiment is proof of Einstein's second postulate?"

"No. There is no proof. The experiment might have triggered an idea, but it's not."

"Until Einstein, how did all other scientists explain the Michelson-Morley experiment's negative result?"

"They don't know. Some think about aether dragging or whatever. But still, it's a negative result, and they couldn't explain it."

"Actually, that is the best answer."

"Beg your pardon?"

"If they don't understand, saying they don't understand is the best answer you can get from scientists. It *is* the most accurate answer – I don't understand. I can't explain, I can't figure it out, I guess it might be this way. I presume I could be wrong, I'm not sure. It states what it is. It's better than an explanation that explains nothing. It's better than an answer that answers nothing."

"You mean Einstein's second postulate doesn't make sense?"

"It doesn't."

"Why?"

"Remember: the postulate consists of three statements. The last statement is nonsense – that is Special Relativity's core problem."

"How?"

"We're getting there. You said before a postulate is something we accept without proof, but can a postulate contradict another postulate?"

"One must give way."

"What about self-contradiction? Can a postulate contradict itself?"

"No."

"Why?"

"Self-contradiction means indecision. It says something, then denies it. It's a non-statement statement."

"Exactly. A theory that allows self-contradiction allows everything. It can be true and wrong, fake and real, alive and dead at the same time. It's a non-theory disguised as a theory. It's non-knowledge disguised as knowledge. It's nonsense disguised as sense. It's irresponsible, useless, and dangerous. Science must not tolerate that."

"But life is full of nonsense."

"That's why we need science to fix it, not endorse it. Who wants to learn the science of nonsense?"

"No one."

"Exactly. To stop nonsense, first, stop self-contradictory knowledge."

"You mean Einstein's second postulate is self-contradictory?"

"Yes."

"How?"

"Let's talk about speed. What does it mean when you say A travels toward B at a speed of 20 km/h?" asked Newton.

"For each hour, A will be 20 km closer to B."

"Hang on. What kind of 'km' and 'hour' are you using?"

"The same as everyone," I said.

"The same as everyone? What if someone else uses a different measurement system?"

"It would be impossible to communicate with him."

"Why?"

"I said bring me a table, he brought a chair. I said meet me before my 9.00 am flight, he came at 11.00 am. I said bring me an apple, he brought me a banana. I said this train's maximum turning speed is 40 km/h, he turned at 100 km/h and the train flew off the rail. I said give a patient a 5 mg tablet, he gave 500 mg, killing the patient. All disasters happen because of miscommunication."

"To avoid that, what should we do now?" asked Newton.

"We must use the same language and same measurement system. Wrong description of the world is dangerous."

"Talking about the same measurement system, you mean Newton's absolute clocks, absolute rulers, absolute measuring devices with absolutely correct readings agreed by everyone. Is that what you mean?"

"Yes."

"Got you. Now, back to the original question: given a perfect world with all perfect measuring devices. If A travels to B at a speed of 20 km/h, you said it means, for each hour, A will be 20 km closer to B until they meet. That is the definition based on Newton's absolute measurement system, right?"

"Yes."

"Does that also mean, for each hour, B is 20 km closer to A?"

"Of course."

"If they start from a distance of 200 km, how long will it take for A to meet B at that speed of 20 km/h?"

"Ten hours."

"What if while A travels toward B at 20 km/h relative to the ground, B also travels toward A at a speed of 30 km/h relative to the ground? How long will it take for A and B to meet?"

"Four hours."

"Why?"

"After four hours, A has traveled 80 km. After four hours, B has traveled 120 km. From the original distance of 200 km, they must meet in four hours."

"From a distance of 200 km, they meet in four hours, so what is the speed of A relative to B then?"

"Fifty km/h."

"So, just a simple addition?"

"Yes."

"Let's presume LA to NY is a straight three million km distance. What if from LA, we send light A toward B in NY at a speed of 300,000 km/s relative to the ground, how long will it take for light A to reach NY?"

"Ten seconds."

"In five seconds, how far has light A traveled?"

"1.5 million km from LA."

"In five seconds, has light A reached NY yet?"

"No. Just halfway."

"I see. Now, let's say that at the same time, we also send light B from NY toward LA at the same speed of 300,000 km/s relative to the ground. How far will light B be in five seconds?"

"Halfway."

"So light A and B will meet halfway in five seconds?"

"Yes."

"From a distance of three million km, light A and light B meet in five seconds, so what is the speed of light A relative to B, by definition?"

"600,000 km/s."

"Does that mean Einstein's second postulate is wrong?"

"I understand it's your calculation based on Newton's absolute measurement system. But it's not the case in reality. The Michelson-Morley experiment tried to find that figure (higher than 300,000 km/s), but it couldn't. All scientists have tried to find that, and they couldn't. Whatever they do, they can only get a maximum speed of 300,000 km/s. That's reality."

"Reality or illusion?"

"What do you mean?"

"If you put two fish into a glass aquarium and from some angle you briefly see five fish, is that reality or illusion?"

"But without vision, how can you tell reality from illusion?"

"What if you empty the aquarium and count?"

"But I cannot empty a Michelson-Morley experiment and count."

"So it's a measurement failure, isn't it?" asked Newton.

"How do you know if it's a measurement failure? We can only say it's a faulty measurement if you know it's different from reality. But in this case, we just don't know what reality is."

"But what reality did Einstein know? What's the speed of light A relative to light B in reality when both travel toward each other, according to Einstein?"

"Still 300,000 km/s."

"How did he know it's reality, given that you said he didn't care of the Michelson-Morley test and never attempted to measure the light speed himself?"

"I don't know. He just said it from his mind."

"He just said it from his mind without any reference to reality?"

"I think so."

"Do you know what it's called?"

"What?"

"Imagination."

"I thought imagination is good. Imagination creates great vision and wonderful ideas."

"But making imagination reality requires hard work and discipline. Not all imaginations succeed. Many have failed and turned into delusion. Special Relativity's second postulate is such one."

"Why?"

"It's a self-contradicting principle."

"Can you prove that?" I asked.

"Yes, we've already done that in the LA to NY thought experiment. Let's dramatize it a bit so you see why it matters. Let's say that during the Earth–Mars war, an alien spaceship lands in Kansas (an empty city between LA and NY, a three-million-km straight distance). According to our intelligence, this last spaceship with one million alien troops onboard will land at 6.00 pm for a few seconds before take-off, and that's the only chance we have to blow it up and finish the war. Each American city has a nuclear bomb hidden underground, which can be triggered by shooting two shots of laser beams (one positively charged, one negative) at the bomb simultaneously from opposite directions. Earth decides to blow up Kansas by ordering LA and NY to shoot a laser beam each toward Kansas at exactly five seconds before 6.00 pm (5.59.55 pm) for the target explosion at 6.00 pm. Both LA and NY have also a million residents each. The question is, which city will blow up and when?" (See Figure 11.1.)

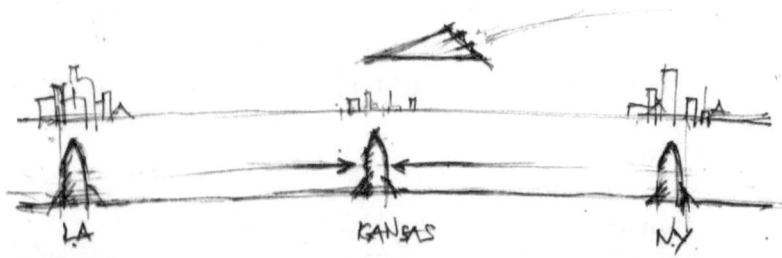

*Figure 11.1 Which city will blow up first, LA, Kansas, or NY?*

"Kansas, of course."

"When?"

"Five seconds later, meaning 6.00 pm sharp."

"From a distance of three million km, if two laser beams reach Kansas in five seconds, what is their speed relative to each other?"

"600,000 km/s." I answered.

"But a relativist would protest, saying that calculation is wrong. According to Special Relativity, the maximum speed of two laser beams relative to each other is still 300,000 km/s. So, from a three-million-km distance, they need 10 seconds to meet. Therefore, to succeed, laser beams must be triggered at 10 seconds, not five seconds, before 6.00 pm (5.59.50 pm)."

"I don't think it works. If we trigger that early, an explosion will occur at 5.59.55 pm, not 6.00 pm. We'll miss the target because aliens would not have arrived by then."

"So what time should we pull the trigger? Five seconds or 10 seconds before 6.00 pm? Which city will blow up, Kansas, LA, or NY?"

"It's so confusing. I think Einstein's second postulate doesn't work. It's untranslatable."

"Why?"

"It gives conflicting results. It cannot be trusted. It's a recipe for disaster. When nothing matters, we can calculate in any way we want. But when it's about life and death, applying that calculation to reality can be fatal. It doesn't save lives."

"What should be the correct calculation?" asked Newton.

"Simple addition."

"Why?"

"It's the definition of speed. If light has a speed, it must add up."

"Does that mean Einstein's second postulate is wrong mathematically?"

"Yes, I would think so. But that is impossible. Math is never wrong." I pondered.

"What if you misinterpret math?"

"How?"

"Come back next week. I'll teach you a lesson about math and the illusion of truth."

"No, please. I'm hopeless with math. No need to teach me. I only want to know how Einstein got that idea about the absolute light speed."

Newton sighed.

"Not sure if I have enough time. Let me try. $1 + 11 = 100$. Is that true or false?"

"False."

"Why?"

"Because $1 + 11 = 12$."

"Keep the same equation. How can you make it true?"

"You can't."

"What if it uses the binary system?"

"Oh yeah, then it's true then."

"Why?"

"Because $1 + 11 = 100$ in binary means $1 + 3 = 4$ in decimal, which is correct."

"What does it tell you then?"

"What's true in this system could be wrong in another system."

"Not quite, but we skip it for the moment. What if I give you this equation: $1 + 3 = 5$, is it true or false?"

"False."

"How can you make it true?"

"You have to create another number system."

"C'mon. Make one."

"Let's say we have a number system as $\{0, 1, 3, 5\}$. That means $1 + 3 = 5$; $1 + 5 = 10$."

"What if I give you another set of equations ($1 + 3 = 5$; and $2 + 3 = 5$)?"

"It's wrong."

"I know. How can you make it true?"

"You can't. No, any number system can give the result 1 = 2. It's a contradiction. It's against math."

"C'mon. Imagine."

"I can't. It's impossible."

"I give you some hints. 1 + 3 = 5; 2 + 3 = 5; 11 = 12; 1 = 2; 251 = 152."

"You mean a number system where 1 can be written as either 1 or 2, like {0, 1(2), 3, 5}?"

"Yes."

"But why would you want to create such a number system – where two different symbols have the same value?"

"But is it true in that system?"

"Yes."

"Isn't that why – that new number system created just to prove 1 = 2?"

"Maybe."


"But does it change the world?" asked Newton.

"No."

"1 ≠ 2, then somehow they make it 1 = 2. Isn't that a huge difference? Isn't that a revolution?"

"It looks like a difference, but it's not."

"What is it?"

"It's an illusion," I said.

"Why?"

"Nothing changes. In another 'weird' language where the symbol '1' and '2' have the same value, it can be written as 1 = 1 and 1 = 2, but it still means 'one' equals 'one,' and not 'one' equals 'two.' Similarly, you can write 1 + 11 = 100 in binary language, but it still means 'one' plus 'three' equals 'four' and not 'one' plus 'eleven' equals 'one hundred.' Different languages have different symbols, but the same meaning. Not a big deal."

"It's not a big deal. But is it misleading and confusing?"

"Yes, it is."

"Why?"

"It causes misunderstanding. When we don't know which system we're using; it's hard to interpret. Like the statement $1 + 11 = 100$, without mentioning the binary system, everyone would say it's wrong because we assume a decimal system."

"Could it be dangerous?" asked Newton.

"Maybe."

"Let's look at this example. Imagine a doctor instructs a nurse to inject a patient with '100 g – binary number – of keraminex' as handwritten in the doctor's medical note. We know '100' in binary means '4' in decimal. Guess what would happen in reality?"

"The nurse injects 100 g instead, and the patient dies."

"Who would be charged with manslaughter?" asked Newton.

"The doctor."

"But the doctor argues that his written note clearly says '100 g – binary number – of keraminex' which means 4 g in decimal."

"But why would he use binary numbers in a medical record? Why didn't he simply write 4 g instead? This is a medical procedure, not an IT test."

"Exactly."

"Why did you tell me this story?"

Newton paused for a few seconds before trying to explain. They were not his best words, though.

"Math speaks different languages. Math is never wrong, but misunderstanding its language can lead to dangerous misapplication with fatal consequences. In binary language, $1 + 11 = 100$ is true; but it doesn't mean 'one' plus 'eleven' is 'one hundred.' Translated into decimal language, that equation means $1 + 3 = 4$. Different languages, different symbols, but same math, and same meaning.

"Special Relativity's second postulate can be written as 300,000 km/s + 30 km/s = 300,000 km/s. The problem is, it's actually written in another language – Einstein's. It might be true in Einstein's language, where km and seconds are not constant. Translated into Newton's language, which is the language we're using for measurement where km and seconds are constant, that equation is written as 300,000 km/s + 30 km/s = 300,030 km/s. Different languages, different symbols, but same math, and same meaning.

"In our scientific language, we use the same words for the same things and different words for different things. Einstein uses the same words for different things and then changes those words' meanings. In Newton's world, one km here is one km there, one second here is one second there. They're all the same units with absolute values. That's why 30 km/s is three times greater than 10 km/s – that expression of difference makes sense. But in Einstein's world, one km here is not one km there, one second here is not one second there; that's why 300,000 km/s here can be different from 300,000 km/s there despite their same symbols. Translated into Newton's world, Einstein's 300,000 km/s could actually mean 400,000 km/s or even 600,000 km/s. Measurement is an extension of language. By altering measurement units, Einstein has altered the language used in Special Relativity. Special Relativity is so relative that it's wide open for any interpretations, including opposite, contradicting, and meaningless ones. You can say it's kind of fun, so long we don't apply it to reality. But when we do, you know what can happen?"

"It can kill," I said, still pondering.

"You look confused. Why?" asked Newton.

"I wonder if we could sum up everything you've said in just three words?"

"I don't have three words; I have three questions. If a car travels toward an old lady at a speed of 100 km/h, but the lady, due to poor vision or miscalculation, thinks its speed is 10 km/h when she crosses the street, what do you think might happen if she got hit?"

"Death."

"A king, last week, asked his accountant to deposit 30 kg of gold into his vault, which already had 300,000 kg of gold. But today, he discovered only 300,000 kg of gold left. When confronted, his accountant blamed it on gold depreciation, value vaporization, or mass curvature. What do you think has actually happened?"

"Theft."

"If two light beams travel toward each other at a speed of 600,000 km/s relative to each other, but a scientist measures only 300,000 km/s, blaming it on slower clocks, time dilation, shorter rulers, longer rulers, contracted space, dilated space, bigger space, smaller space, space–time curvature, different reference frames, or whatever, what do you think of that explanation?"

"Fake."

*Figure 11.2 Two cars*

"I have a billion-dollar challenge for you," said Newton. (See Figure 11.2.)

"A billion dollars?"

"Yes, for any student to break Relativity's second postulate."

"I'm listening."

"If two cars start at the same time traveling at the same speed toward each other from opposite ends of the road, where would they meet?" asked Newton.

"Halfway."

"Always?"

"Yes, always. Regardless of the speed or the road length."

"What if two light beams travel at the speed c (300,000 km/s) relative to the ground?"

"Same, halfway."

"What if the road is 2 'c' (600,000 km) long?"

"Same, halfway."

"How far is halfway?"

"Same value as 'c' (300,000 km)."

"How long does it take for two beams to meet halfway?"

"One second."

"From a distance of 2 'c,' if two beams can meet in one second, by definition of speed, what is their relative speed to each other?"

"2 c."

"Does that contradict Relativity's second postulate?"

"Yes."

"Which one is wrong?"

"I'm confused, sir."

"Did you pass Year 7 physics?"

"Yes."

"Then you must know."

"I suspect the second postulate."

"Here is the challenge," Newton continued. "Can you design and conduct a real test, not a thought experiment, to convince all physicists from your world that Relativity's second postulate formulated in its velocity addition formula is absolute nonsense?"

"Impossible. It's a monumental postulate on which Relativity has firmly stood for over a century."

"That's why I offer one billion dollars for any student to break it. Can you?"

# CHAPTER 12
# CLOCKS

*If the effect is an actual change in the rate of an atomic clock —
which is quite a rational idea — it cannot be extended to the time
duration of other processes, such as the ageing of an individual.*
                                                    *LOUIS ESSEN*

*Too many physicists subscribe to the belief that there is nothing
to time except what can be seen on the face of a clock;
but that amounts to the ridiculous statement
that a measuring device has been built to measure nothing but itself.*
                                              *JOHN E. CHAPPELL, JR.*

"Can we talk about clocks now, sir?"

Finally, we met to discuss the topic he'd delayed for a while.

"What do you want to know about clocks?"

"How are they made?" I asked.

"What kind of clocks do you want?" replied Newton.

"Clocks that show correct time."

Newton said, "Tell me more. Define clocks."

"Time telling devices."

"Tell me more. Can this world survive without clocks?"

"No."

"Why?"

"Without clocks, we would get up and not know what time to go to work. Buses wouldn't know what time to run. Rail crossings wouldn't know when to shut down the gates to avoid accidents. Passengers wouldn't know what time to board airplanes. Planes couldn't take off without knowing what time other planes would land. Factories couldn't start because half their workers hadn't turned up. Workers couldn't get paid because employers knew nothing about their working hours. Scientists couldn't measure speed without clocks. Speedometers won't work without clocks, so we wouldn't know what speed a superfast train was turning at a curve. Traffic lights won't work without timers or clocks. We couldn't have elections because we wouldn't know what time voting would start. Football games would run forever because no one would know what time to end them. Olympic games would be canceled without clocks for measurement. Competitions would be canceled because no one would know the start time. Students wouldn't know what time exams would start or end. Without clocks, all synchronized or simultaneous group activities would be suspended. The whole world would collapse."

"So you want simultaneity?"

"Yes."

"Your world survives on simultaneity. Simultaneity is defined as events happening at the same time. Without clocks displaying the same time, simultaneity is impossible. Is that right?"

"Yes."

"And that is the purpose of clocks – clocks must always show the same time?"

"Yes."

"I see. To measure change, we need something that is always the same. It can be challenging in a world where everything constantly changes, but that's the purpose. It's the very first principle of measurement. All rulers, scales, thermometers, speedometers, odometers, and gravimeters must have the same reading. Otherwise, if a length is 10 cm according to this ruler but 8 cm according to another ruler, you wouldn't know which one is true and which is false. Agree?"

"Yes."

"So apart from the same display, what else do clocks need to have?" asked Newton.

"That's all."

"Are you sure?"

"Yes, same time display. That's all I ask."


"Do clocks need to have a constant speed? Does the time-lapse between hours need to be the same, say between 1.00 pm to 2.00 pm and 5.00 pm to 6.00 pm?" Newton asked.

"Of course," I said.

"What for? You said before that the same-time display is all you need, right?"

"I've just realized I need more. Because if, say, 1.00 pm to 2.00 pm is longer than 4.00 pm to 5.00 pm, all workers would prefer working late. Why would they want to work longer hours for the same pay?"

"I see. For fair work."

"Not only that, accuracy matters. I mean, a man normally can dive for a maximum of three minutes (and may die at five minutes), but if he dives between 1.00 pm and 2.00 pm, three-minute-diving actually would last longer and might kill him."

"I see. That's why the same time-lapse is another condition clocks must have. In short, all clocks' speeds must be the same and constant. Agree?" asked Newton.

"Yes. Same and constant."

"Anything else?"

"That's it. Two conditions are enough."


"What about lifespan? Do you want a perpetual clock that runs forever?"

"Yes, it doesn't hurt to have a perpetual clock."

"It doesn't hurt, but it's not free."

"What's the price?"

"It depends on what you mean by perpetual. Let's say, 100 years is one million dollars. One hundred thousand years is one billion dollars. You want more?"

"No. How about two months?"

"Are you sure you don't want a perpetual clock?"

"It's out of my budget. I only need two months for testing, that's all."

"Well, that'll surely reduce the price. What else do you need? Do you need it to be waterproof?"

"What for?"

"It's your choice. You're a customer; you're the boss. The more I know about your need, the better I can serve."

"Ok. I'll have it waterproof as well."

"Ok. How much is waterproof? One km depth in water or more?"

"Does it affect the price?"

"Of course. Nothing is free."

"Then I won't have waterproof."

"Your call."

"But what if I accidentally drop the watch into water for five seconds?"

"Out of warranty, it might stop working."

"It's so confusing. All I want is a watch that has the correct time. But now I realize there are so many other factors."

"That's life. Welcome to the real world. What you want and what you need are two different things. We don't often realize that until we face reality. That's why we ask – so we better understand."

"How about giving me a catalog with a price list, so I can choose?"

"There you are," Newton said, passing a blank sheet of paper.

I looked at it, saw nothing, and nodded.

"What is the cheapest clock you have?"

"A hundred dollars."

"And the most expensive one?"

"A hundred billion dollars, and counting."

"That must be a perpetual clock?"

"Close. It's guaranteed to run for 100 billion years – life warranty certificate provided." Newton grinned.

"I wouldn't buy it." I was skeptical.

"I know."

"What is the same thing between all these clocks?"

"Same time display."

"What's the difference between them?"

"Price tags and working conditions."

"What are the working conditions for the 100-billion-dollar clock?"

"Nothing. You do nothing to make it work. It's already perfect, unconditional, unlimited. It's waterproof, heatproof, force-proof, vibration-proof, gravity-proof, damage-proof, explosion-proof, error-proof, fault-proof, and change-proof. You can try to blow it up with a nuclear bomb, and it will survive."

"What are the working conditions for the 100-dollar clock?"

"A lot. You have to do a lot to make it work."

"Like what?"

"Don't put it in water because it's not waterproof. Don't shake it because it's not vibration-proof. Don't put in extreme heat because it's not heatproof. Don't put it in a freezer because it's not ice-proof. Don't change its gravity because it's not gravity-proof. And most importantly, don't manually wind up its hands, because that'll surely change its time." Newton grinned again.

"Why would I do that?"

"Some people don't know what they're doing, and they keep asking why."

"Can it work correctly in inertial motion?"

"Yes, by default, all my clocks, cheap or expensive, are all inertial-motion-proof."

"Why?"

"Because the laws of physics remain the same in inertial motions. Objects' behavior or their speed won't change, according to the first postulate."

"Can I use your 100-dollar clock in space travel?"

"Yes, only if it survives the take-off and all associated changes in gravity, acceleration, g-force, atmosphere, zero-gravity, heat, and so on."

"You mean those changes can impact the clock and make it faulty?"

"Yes."

"Can I adjust its time once it becomes faulty?"

"I'm not sure, but you're welcome to test. After take-off and as soon as you're in inertial motion, if it survives, you can try to sync your clock to any other correct clocks, and see if it works."

"What about the clock's battery? Would it be affected during take-off?"

"Why worry? You just take extra batteries and change when required."

"I'll take the cheap clock, then."

"I think it's a good choice, not because it's cheap, but because it's good enough for your purpose. Instead of having a perfect clock, why not have an imperfect one that still works within its working conditions? It's much simpler."

"I see."


"Now you know what you want with clocks; let's see if you can make them. Imagine you're a clock-maker. Would you build clocks to have the same speed?" asked Newton.

"Yes, of course."

"No, not necessarily. Clocks don't need to have the same speed to show the same time," Newton reasoned.

"How come? You've previously said all clocks must have the same speed."

"Let me clarify. Same speed for conceptual time, but different speed for physical parts."

"What do you mean?"

"I have 21 clocks here. Have a look and tell me if they have the same physical speed."

I thought I'd seen them all. He had many weird clocks with different designs. Pendulum, sand-clocks, digital, etc. Initially, I thought all his analog clocks displayed different times for different cities around the world, for their hour hands were pointing in different directions. But after a careful look, I realized they all showed the same time (7.10 pm) in Melbourne, but their hour hands were pointing to different positions. Some analog clocks had only one hand indicating hours. Some had 24 numbers. Some had only six. And that meant their physical hour-hands had different speeds. (See Figure 12.1.)

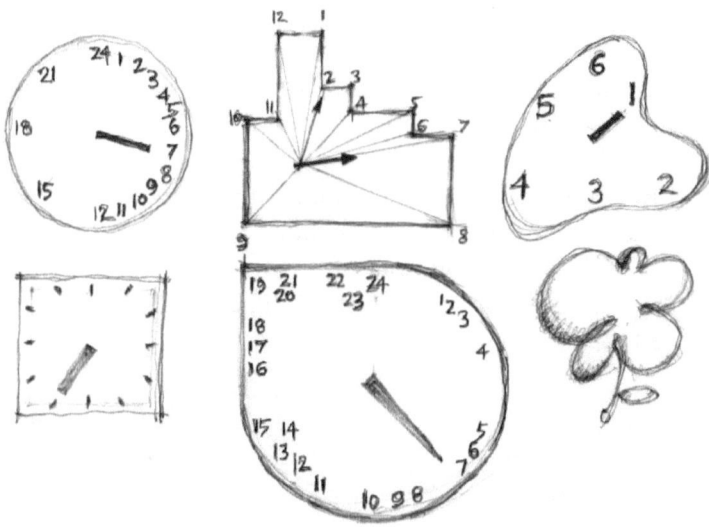

*Figure 12.1 Different clocks. Same time.*

"Did you make these clocks?"
"Some."
"Why did you make them so different?"
"So that you can see clocks with different speeds can still show the same time. You just readjust their markings accordingly."

"I see."

"What more do you think clocks need?"

"Each minute must have the same value, so the clocks' speed must be constant."

"No, you're wrong."

"Where am I wrong? The constant time or the constant speed?"

"You're right about constant time – each minute must have the same time-lapse. But to achieve that constant time, a clock doesn't need a constant speed. Look at the purple clock on the wall. It has only one hand – an hour hand, 24 numbers, and irregular gaps between those numbers. That means the hand needs to travel faster during long gaps and slower during short gaps to display the same conceptual time as other standard clocks. Again, you just rearrange its physical marking accordingly."

"Weird. So all these clocks still display the correct time despite their different speeds?"

"Yes."

"But not this one?" I pointed to a green-colored clock near the kitchen with only one hand.

"Why?" asked Newton.

"Its hand doesn't move at all. The clock's dead." (See Figure 12.2.)

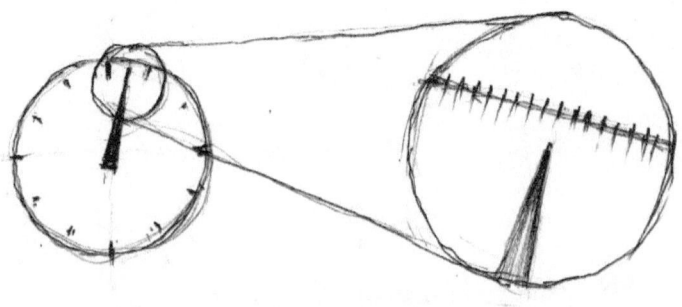

*Figure 12.2 An accurate "frozen" clock*

"No, it's still working," Newton corrected.

"Why? For all the time I've been coming here, its hand always points to something like number 1."

"It looks like a frozen clock with frozen time, but it's actually working. The tip of its hour hand moves extremely slowly, about one mm yearly. Its hand has taken almost 10 years to move up that first mark. It would complete its full cycle in 120 years."

"How can you read the hour?"

"You need a super microscope to see."

"Weird."

"Clocks can stop, time never."

"I see."


"Do clocks need the same design and mechanism?" asked Newton.

"No."

"You're right, they don't. We have digital, analog, pendulum, sand clocks, water clocks, color clocks..."

"What do you mean by color clocks?"

"Clocks without hands or numbers."

"How can you read them?"

"You don't read – you see. They tell time by illuminating different colors at different times."

"When is 6.00 pm?"

"Blue."

"7.00 pm?"

"Purple."

"7.10 pm?"

"Fuchsia, kind of red-purple."

"7.12 pm?"

"Magenta."

"I cannot tell the difference."

"You can't. But artists can, like Picasso, Monet, or Van Gogh."

"Why did you make such a clock?"

"Time doesn't stop because of a faulty clock or because you can't read its time."

"I see." (See Figure 12.3.)

*Figure 12.3 Color clock*

"What I'm trying to show you is that time can be conceptually represented in many different ways. It's about how we look at change. As a matter of fact, almost any change whose speed we already know, calculate, and correctly predict can be used as clocks to tell time. You understand?"

"Yes. But that only works for change with a clearly different output?"

"Of course. If the output is always the same, you can't read its time. Like color clocks, if you can't recognize the difference in color due to color blindness, you can't read the time. It's useless for you. But it's useful for artists with good eyes."

"What if I wear sunglasses or different-colored glasses?" I asked.

"Good question. What would you see then?"

"Different color."

"How do you read?"

"Different time."

"What does that mean?"

"The clock becomes faulty."

"Is the clock faulty, or is your reading faulty due to different lighting?"

"Whatever the cause, it's still faulty, according to me."

"What happens to other digital clocks?"

"They're still correct."

"What does that mean?"

"One clock is faulty, but other clocks are still correct."

"You're absolutely right."

"But what happens to other people who see that color clock with their naked eyes without glasses?" I asked.

"What would they see on that color clock then?" Newton asked me back.

"Still correct time."

"What does it tell you then?"

"The same clock is faulty to me but correct to others."

"What does it tell you then?"

"We have different eyes."

"What does it tell you then?"

"Vision can be an illusion."

"Exactly. Time is absolute, but time visibility can be relative due to illusion. Talking about illusion and faulty clocks, let me show you something."

Newton pointed at a clock on the other side of his kitchen. It was the weirdest clock I've seen. It had 10 digits, which were randomly placed, unevenly apart, with counter-clockwise hands. A clock with reverse time, random time, wrong time – it was impossible to read. (See Figure 12.4.)

"What the hell is that?" I asked.

"What do you see?"

"A faulty clock."

"Call it a faulty clock, fake clock, crazy clock, wrong clock, or whatever name you like, but actually, it's a correct clock."

"Why?"

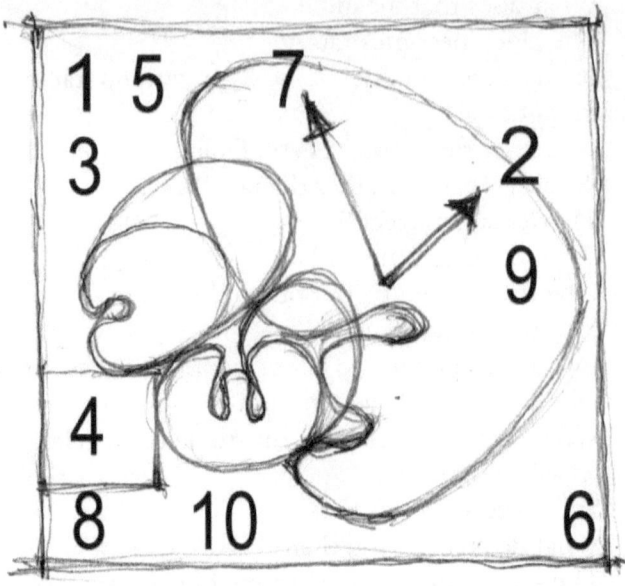

*Figure 12.4 If you don't know what time is, how can you tell time dilation?*

"You cannot read it because visually it's wrong."

"Then how can you read?"

"You don't read – you hear. When you press the button on its back, it will read out the correct time for you. The clock has two mechanisms producing two different speeds. One tells the correct time; the other tells the wrong time by tricking sighted people. It's a clock for blind people who cannot see and cannot be tricked visually."

"I see."

Newton shrugged, "What you don't see will hit you harder than what you see."

"Is that why you made it?" I asked again.

"Yes. Everything we make has a purpose. Without a clear purpose, things we make can fall apart. Time is just a conceptual speed – physically displayed by one of millions of different speeds inside a clock. Choose one to represent time, but don't think that physical representation would last forever. Also, whatever happens to that representation, other speeds might not be affected. When a clock runs out of battery or outside its working environment, it can stop working. But even when a clock stops, other speeds are still running. Speed of neutrons spinning around their atom. Speed of corrosion on different parts of the clock. Speed of paint fading away. Light reflecting from the clock. Dust covering on the clock surface. The air circulating inside the clock. Some viruses still moving inside the clock. And those plastic parts slowly breaking up at the same rate."

"One speed stops; life goes on."

"Absolutely."

"I think I know what you mean. You try to say everything in the universe has a speed. Speed matters. That's why we need to measure speed by first choosing a standard speed as a measurement unit, and that representation is called time."

"You're absolutely right. Speed matters. Dimension matters. Temperature matters. Mass matters. Economy matters. Finance matters. For anything that matters, we need to name and measure it. To measure, first, we need to imagine a measurement system with absolute units that don't change. The reason 10 km/h is slower than 12 km/h is because they both use the same measurement unit 'km/h,' which is the same value anywhere. Otherwise, nothing makes sense. Measurement is an extension of language. To describe the world correctly in science, we need both correct language and accurate measurement."

"And clocks – as time representation – are to provide that measurement."

"Exactly."

"For clocks to work, they must have a certain speed?" I reasoned.

"Yes."

"Since clocks are manmade, we can design clocks to have any speed we want?"

"Yes."

"Since clocks are manmade, we can make correct clocks displaying correct time, or faulty clocks displaying faulty time?"

"Yes."

"Since clocks are manmade, we can make clocks run at different speeds, freeze indefinitely, run backward, or display the wrong time?"

"Yes."

"Does that also mean we can make clocks run according to Einstein's time dilation equation if we want?"

"Yes, you're absolutely right. Devices are manmade. You can make them any way you want. You can make either correct devices or faulty devices. You can deliberately use a faulty scale to claim quick weight loss. You can use a faulty emission lab test to claim the cleanest diesel car. You can use a faulty clock to claim time dilation. If you want to cheat by using faulty devices, surely you can. But if you want correct devices with correct measurements to prevent disasters, you can, too. So what do you want?"

"I don't want to cheat. I don't want to confuse people. I want correct clocks with the correct time. How can I make perfect clocks?" I asked again, the second time.

"Do you drive?" asked Newton.

"Yes."

"Do you have a car?"

"Yes."

"You might not know how to make a car, but do you know how to keep your car in good shape?"

"I have it serviced regularly. I look after it. I wash it every week."

"Do you drive it without a wheel?"

"No."

"Do you put water into its petrol tank and drive it?"

"No."

"Do you deliberately crash the car into a wall at high speed?"

"No."

"So you don't do anything silly that you know would damage your car, right?"

"No."

"Making a perfect clock is like making a perfect car. You might not know how to make it, but you must not do what you know would break it. Do you think you should take a hammer and smash your clock?"

"No."

"Do you think you should open the clock cover, pour petrol on it, and set it on fire?"

"No."

"If you know your clock is not waterproof, would you place it in the water?"

"No."

"If you know your clock might get damaged in a certain condition, would you put it in that condition?"

"No."

"If you know your atomic clock doesn't work well in different gravity, would you put it there?"

"But why doesn't it work well in different gravity?"

"Cesium density increase, for instance."

"Beg your pardon?"

"Do you know how atomic clocks work?"

"No."

"Don't you think you should ask those clock makers that question before use?"

"Yes."

"If they tell you atomic clocks can give different readings at different gravity, would you listen to their advice?"

"Yes. But no one has told me."

"Whose fault is it?"

"Not my fault. I'm a user. If there's something users should not do, those clock makers must warn me."

"So you're a naive user?"

"Innocent user."

"I see. You are what you are. But if you're a clock-maker, would you tell your customers?"

131

"Of course, I would if I knew."

"As a clock-maker knowing how to make a clock, would you know how to break it?"

"Yes."

"Why?"

"I mean, if I know how to make a car run, I should know what to stop the car run. If it's an EV, I'd empty its battery. If it's an airplane, I'd turn off its engine. If it's a submarine, I'd remove its propeller. If it's a spaceship, I'd detach its booster. The same applies to trains, buildings, submarines, missiles, or clocks; if I can make them, I can break them. You just do the opposite of what works."

"That means you already know the answer then."

"No. I still don't know how to make perfect clocks. Knowing what doesn't work doesn't mean knowing what works. It helps narrow your search, though."

Newton seemed hesitant.

"What is your question, again?"

"How to make perfect clocks?" I asked the third time.

Newton paused. He looked through the window to the sky outside, saying nothing. I was clueless about what happened. It's a simple question. Why can't he answer? Something is not right. I suspected he struggled. Does anyone struggle to create something that works forever? Or is it a wrong question? After a few minutes, he replied. "You want a short answer or a long one?"

"Short one, please."

"I don't know."

"Long one, please."

"Life is a clock. What you see in a clock is everything you want in life. You want control. You want order. You want certainty. You want everything to happen the way you want, the way you predict, the way you expect. The problem is it's human desire. Reality always has surprises. So does a clock.

"Do you want to make a perfect clock? It's the same thing as you want a perfect marriage, perfect relationship, perfect job, perfect country, perfect machine, perfect body, perfect mind, or perfect car. How can you make a perfect car that never fails? How can you make a true equation that fits all universes? How can you find a perfect partner who always loves you? How can you make a perfect machine that never breaks? How can you find a perfect game that you always win? How can you make a perfect world free of poverty, war, and violence? How can you find happiness that lasts forever? Simple question, but no answer. A clock, the closest thing to life that man can ever control, is no exception.

"I don't have the perfect answer, but I do have some suggestions. If you make a plane that cannot fly, do you know it's faulty? If you make a car that cannot run, do you know it's faulty? If you make a ruler that shrinks, do you know it's faulty? If you make a clock that shows different times from other correct ones, do you know it's faulty? Why waste time finding a perfect device in billions of years if you cannot even tell a faulty one right now right here in front of you?

"And that's a problem with this world. You ask the wrong question and try to answer it. You ask for something you don't fully understand, then you rush to find the answer. You end up getting lost because you don't know what you're doing and where you're going. Asking is easy, but asking right is not.

"Let's fix the question first. It's okay if you don't have a perfect solution that works forever; try something that works today. If your clock stops running in water, then don't put it in the water, or use a waterproof clock instead. If your clock stops running in extreme heat, then don't put it in extreme heat, or use a heatproof clock instead. If your partner stops loving you because you cheat, then don't cheat or find a cheat-proof partner instead. And if your atomic clock gets faulty because of different gravity, don't put it in different gravity, or what should you do?"

"Use a gravity-proof clock instead."

"Exactly. I might sound complicated, but I hope what you find is simple. Try not to make a perfect clock that lasts forever; ask yourself if you can recognize a faulty clock right now or not. If you don't know, you don't know. But let's focus on what you know. If you know a clock is not faulty for the moment, use it. If you know it's faulty, know what you should do?"

"Fix the clock, not time."

# CHAPTER 13
## 100 YEARS OF FAKE EVIDENCE

*The Hafele & Keating experiment may well rate
as one of the biggest hoaxes in the history of modern Science.*
AL KELLY

*Einstein's use of a thought experiment,
together with his ignorance of experimental techniques,
gave a result which fooled himself and generations of scientists.*
LOUIS ESSEN

*There is not a shred of real evidence in favor of time dilation,
as opposed to clock retardation.*
JOHN E. CHAPPELL, JR.

*It is difficult to get a man to understand something,
when his salary depends upon his not understanding it.*
UPTON SINCLAIR

*When a person's livelihood is at stake, they tend to not rock the boat.
If you are a particle physicist who knows Einstein's increase in mass is
not real, you don't go out and tell the world this
if you want to keep your job.*
DAVID DE HILSTER

*That scarcity of resources, and the pressure for academics to regularly
publish in order to secure further funding, could be one reason*

*why an increasing number of researchers*
*have been caught cheating in recent years.*
ELISE WORTHINGTON
*ABC investigation journalist*

## The Hafele-Keating Experiment

"I was about to ask you about the Hafele-Keating experiment…"

"What about it?" asked Newton.

"But I don't think I need it anymore."

"Why?"

"I got the answer. I know the secret of clocks now. There is no evidence of time dilation, only faulty clocks. Never have, never will."

"Since you know that secret, do you find the Hafele-Keating experiment one of the world's most stupid scientific tests in mankind's history?"

"Beg your pardon?"

"Sorry, I'm wrong. It's not one of the worst. It *is* the world's most stupid scientific test in mankind's history. And if your physicists cannot see it, there is no hope for physics."

"I don't understand. Why would you say that?"

"There are millions of imperfect devices which will become faulty one day; you don't need to carry out another test just to prove they can be faulty."

"The test is meant to prove time dilation, not faulty clocks."

"But would you test the devices first before use?"

"The Hafele-Keating test was the first, and those clocks failed immediately."

"So the Hafele-Keating test has proved that atomic clocks can be faulty?"

"Yes."

"Why would you waste money flying atomic clocks around the world just to prove they can be faulty? Why not give me a ten-dollar hammer – I'll smash those clocks and tell you if they are faulty or not."

"Not by force. They wanted to test if inertial motion can change clocks."

"But Einstein's first postulate states inertial motion cannot cause any change. So do they want to prove or disprove Einstein's theory?"

"Yeah, I think the test itself is confusing. They say inertial motion cannot change objects' behaviors. Yet they wanted to prove inertial motion can cause time dilation, which (they believed) can be visually seen on clocks – meaning change to objects' behaviors. That's a double standard. That's a self-contradiction. Also, the Hafele-Keating test was carried out, not as a strictly inertial motion, but as a circular motion with other interferences, such as acceleration, force, vibration, gravity, etc."

"So the test itself is nonsense?"

"Yeah, I guess you could say that."

"What about physicists who claim they have evidence of time dilation?"

"I think they're confused."

"Why?"

"I don't know."

"Can you possibly prove heaven is bad?" Suddenly, Newton changed direction.

"You can't. Heaven is good by definition." I protested.

"Can you build heaven on Earth?"

"We can try."

"Let's say we've built one, such a perfect castle that everyone comes, enjoys, and loves so much that they name it heaven. But after a few years, say one day, flash flooding damages its sewerage system, pushing tourists away. What would you call it then – that the castle is no longer heaven or heaven is bad?"

"The castle is no longer heaven."

"Exactly. Time is like heaven. If heaven is never bad, time never dilates. To visually see heaven, we need to create a physical castle. To visually see time, we create a physical clock. As long as the castle is perfect, it represents heaven. As long as the clock is perfect, it represents time. If one day the castle collapses, we'd call that a faulty castle, not bad heaven. If one day that clock delays, we'd call that a faulty clock, not time dilation."

"Time dilation doesn't exist, only faulty clocks."

"Exactly. You previously said they're confused – those physicists who use faulty clocks as evidence of time dilation. What confuses them?"

"I don't know."

"Think."

"Their language."

## Space–time curvature

"What about space–time curvature?" I asked.

"Have you ever seen a car compressed into a slab of metal scrap one-tenth of its original size?"

"Yes."

"Is it evidence of what?"

"Physical object deform."

"Is it evidence of space–time curvature?"

"No."

"If there was evidence of space–time curvature, do you think physicists would show you?"

"Yes."

"But if you could visually see it, it must be physical, isn't it?"

"Yes."

"And if that is physical, what would you call it?"

"A physical object deformed, curved, bent, but not space–time curvature."

"So where is the evidence of space–time curvature?" (See Figure 13.1.)

"But is it true that mass causes space–time curvature?" I asked.

"That is a wrong question. Asking wrong questions and answering them is a waste of time. What is space–time curvature? That's the question you should ask before asking what is the cause of space–time curvature. If there is no murder, how can you find a murderer? If you cannot define ghosts, how can you tell where to find them? If there is no time dilation, how can you talk about the cause of time dilation?"

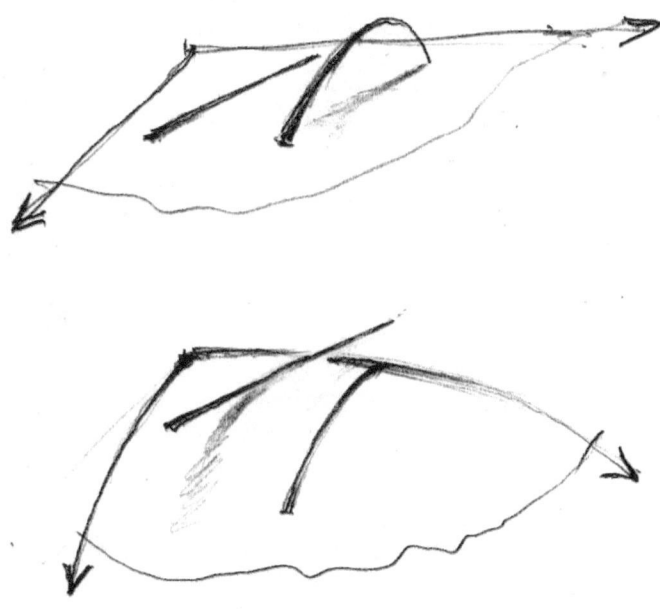

*Figure 13.1 Changing languages doesn't change ideas, only expressions*

"What is space–time curvature?"

"Who created that term?"

"Einstein."

"Why don't you ask him?"

"I wish he was here."

"What about other mainstream physicists? What do they mean by space–time curvature?"

"I'm not really sure. They use the term without a clear definition. Maybe they don't mean it physically, but conceptually."

"How can you possibly have physical evidence of a conceptual space–time curvature?"

"You can't."

"But conceptually, what does space–time curvature mean?"

"No idea. They talk about space combined with time as a 4D space–time; mass curves space–time; planets circle the Sun because the Sun warps space–time and so planets travel in a straight line in 4D but seen as a curve in 3D, etc." (See Figure 13.2.)

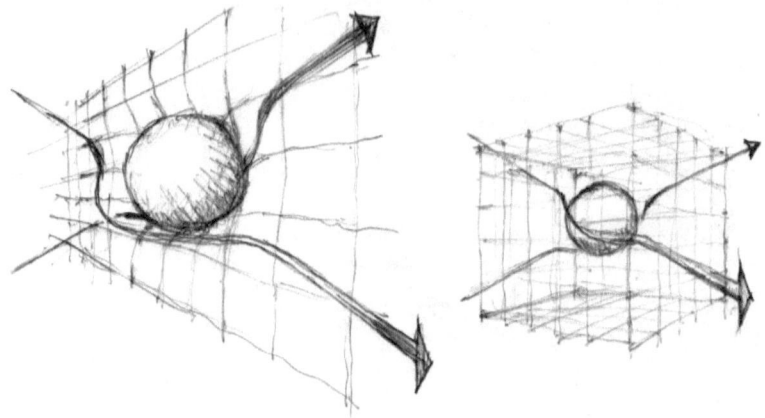

*Figure 13.2 If space curves around an object, where is that object located?*

"Are you talking about physics or math?"

"Physics deals only with physical things. So it must be math, then."

"In Math, can a straight line in 3D ever be seen as a curve in 2D?"

"No."

"Then why would a straight line in 4D be seen as a curve in 3D?"

"That's why I don't understand."

"Let me teach you something about higher-dimension geometry. It's nonsense to say a straight line in 4D is seen as a curve in 3D. A 4D straight line, when viewed from 3D space, may be invisible, partly visible as a point, fully visible as a straight line, but never visible as a curve. A curve in 3D will always be visible as a curve in 4D or any higher-dimension space. (See Figures 13.3, 13.4, and 13.5.)

*Figure 13.3 1D curvature? Or a curved object in 2D space?*

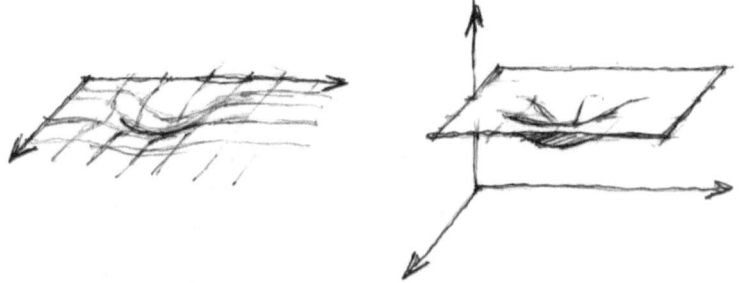

*Figure 13.4 2D curvature? Or a curved object in 3D space?*

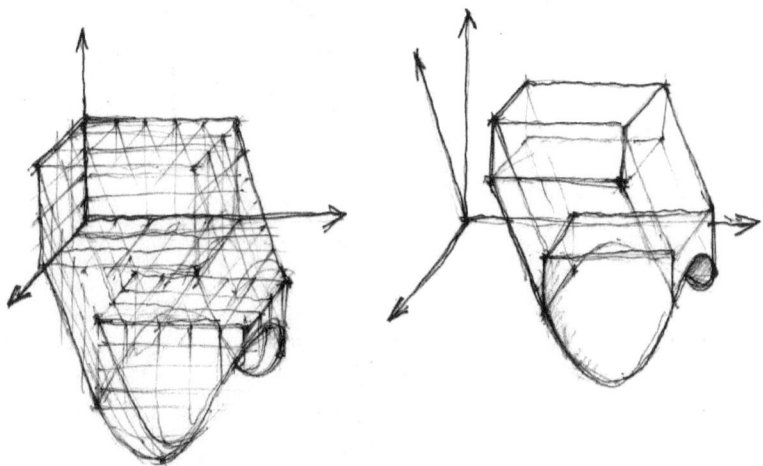

*Figure 13.5 3D curvature? Or a curved object in 4D space?*

"In fact, any change to space dimensions doesn't impact an object's nature, but only its visibility. When its space dimensions increase, an object is always fully visible; when it decreases, the object might be fully visible, partly visible, or completely invisible.

"Space–time to the universe is like longitudes and altitudes to Earth. Longitudes don't self-exist. We create them. They're like a coordinate system for navigation. If 2D is enough, 3D space is not required. If 3D is not enough, we use 4D. And so on with 5D, 6D, 7D, and more. Higher-dimension space is only required for something invisible or unidentified in a lower dimension. (See Figure 13.6.)

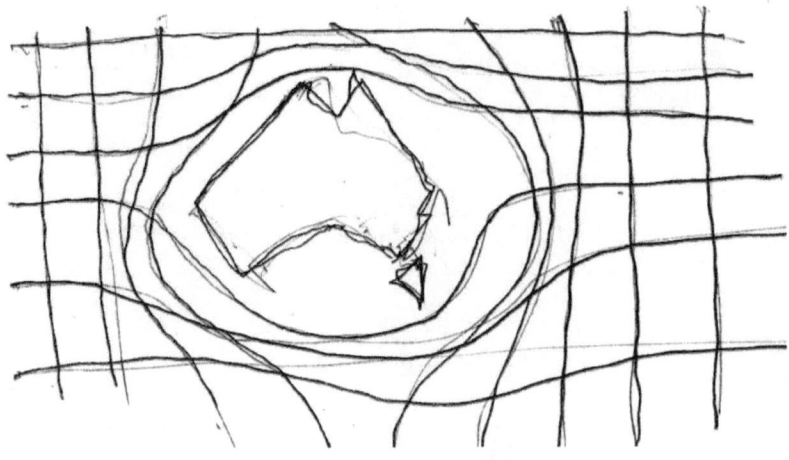

*Figure 13.6 If we use curved longitudes and altitudes, where is Australia?*

"But it's ridiculous to say something exists only in the fourth dimension when it's fully visible in 3D. If something is fully visible in 3D, that thing cannot exist exclusively in the fourth or any higher dimension.

"Time and space are not real. They're not physical, and cannot be bent, dilated, contracted, curved, modified, or interfered with. As an imagination product, space can have any dimensions we want – 2D, 3D, or 10D – depending on how many variables a point is defined as different from another. Space–time is a reference system we create to index anything that matters to us. (See Figure 13.7.)

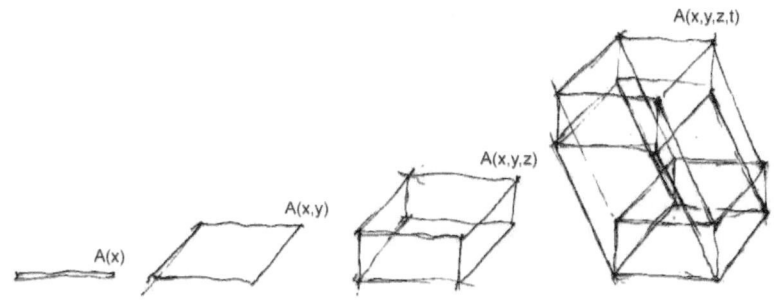

*Figure 13.7 Illustrations of 1D, 2D, 3D, and 4D on 2D paper*

"Like time dilation, space–time curvature is a misconception. Mainstream physicists have confused speed with time and object with space. Instead of changing speed, they change time. Instead of changing objects, they change space. Then they use evidence of speed change to prove time dilation and deformed objects to prove space–time curvature."

## Muons

"What about muons?" I asked.

"What about muons?" He asked again.

"They say muons decay differently when traveling in space than on Earth. And they use that as evidence of time dilation."

"How can you explain that?"

"Change to their decay speed due to gravity, heat, accelerations, force, etc. Almost every change in the universe can be described as speed change – meaning the end of time dilation. And as long as Einstein's first postulate confirms the laws of physics remain the same in inertial motion, no change can happen due to inertial motion, but to force, gravity, heat, acceleration, and so on."

"You say it all."

## GPS

"What about GPS?"

"What about GPS?"

"They say GPS is proof of time dilation."

"What do you say?"

"It's proof of imperfect GPS clocks. Their clocks need adjustment or re-sync to work correctly."

"What makes GPS clocks faulty?"

"Acceleration, g-force, gravity change during launch, maybe."

"What else?"

"There is other time dilation evidence I wanted to ask you, but now I realize there is no need."

"Why?"

"There is no evidence of time dilation – never. Just speed change. Nothing else."

"Fine."


## Simultaneity

"Can I ask you something else – about simultaneity in Special Relativity?"

"You mean evidence of relative simultaneity?"

"Yeah, I don't understand the fuss about simultaneity, but it seems a big deal in Special Relativity to many people. Do you think Einstein's story about train and lightning events is evidence of relative simultaneity?"

"No."

"What is it?"

"It's evidence of language confusion. Nothing else. One event has thousands of mini-events inside. Each mini-event has thousands of micro-events inside. We don't know what event we're talking about. We use the same word 'event' for many 'different events' in the story. Since each event occurs at a different time, we mistakenly think one single event can happen at different times, and that's the basis of relative simultaneity, which is nothing but a misconception."

"Can you please explain?"

"Imagine one night you look up and see the Moon and Jupiter blowing up at the same time. Can you say those two events are happening at the same time?"

"Which event?" (See Figure 13.8.)

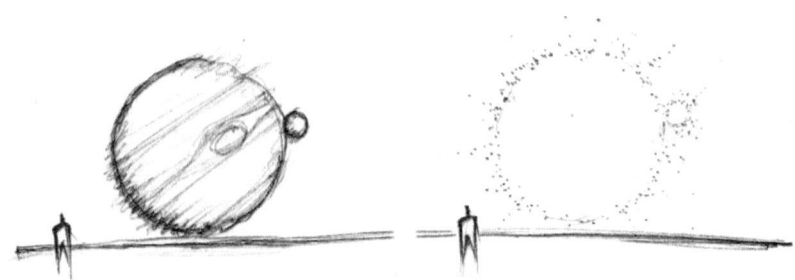

*Figure 13.8 The disappearance of Jupiter and our moon*

"Good question. Let's say the moon explosion event is when the moon is blowing up. Since the moon is about 380,000 km from Earth, it would take over one second for people on Earth to receive that image, which we call the Moon-explosion-observed-from-Earth event. Similarly, we call Jupiter-explosion an event when Jupiter blows up and Jupiter-explosion-observed-from-Earth an event when that image reaches Earth 40 minutes later. Are those two images observed at the same time from Earth?"

"Yes. Both explosions' images reach Earth at the same time. I can see them happening at the same time."

"Can you take a photo of those two events happening in Earth's sky simultaneously?"

"Yes."

"Good. But do those explosions really happen at the same time?"

"No. The Moon is closer to Earth than Jupiter. So Jupiter's explosion must have happened long before the Moon's to have their images reach Earth at the same time."

"What if you stand near Jupiter? Which explosion would you see first (if you survive)?"

"Jupiter's explosion first, then the Moon's."

"Can you take a photo that shows Jupiter blowing up while the Moon is still intact?"

"Yes."

"What if you stand near the Moon? Which explosion would you see first (if you survive)?" asked Newton.

"Since you can see both explosions in Earth's sky, it means Jupiter's explosion has happened 40 minutes earlier than the moon. So, if I stand next to the Moon right at the moment the Moon is blowing up, the image of Jupiter's explosion must have reached the Moon already. That means either I might see both explosions, or Jupiter's image might have just disappeared by then."

"From three photos, what can you tell about simultaneity? Which one happens first, which one later, or do they both happen at the same time?"

"You can't. Photos can be deceptive. You can only see images of events, but not original events. So unless both events happen at the same place, you cannot tell if they happen simultaneously or not, even though their images happen simultaneously to some observers."

"What if you have absolute clocks with absolute time?"

"That would be perfect. That's exactly why we need absolute time. We will define simultaneity as events that happen at the same absolute time regardless of their locations. That said, whether their images can be seen simultaneously from another distance is another story."

"In short, posting a letter and receiving it are two different events?"

"Yes," I replied.

"You see the importance of having absolute clocks with absolute time?"

"Yes, to stop confusion."

"And the importance of correct description for each event?"

"Yes, to stop misunderstanding."

"Confusion about time and events leads to confusion of simultaneity. Einstein's relative time creates that problem. Newton's absolute time corrects that."

## Illusion of Length Contraction in Motion

Newton continued: "That also raises another issue with length contraction. It's tricky to measure something moving at high speed based on its image. We want to measure the object's length, but we mistakenly measure its moving image's length. If they're the same, it's okay. But if they are not, we'll end up measuring the wrong thing. Two different things give two different measurements. But since we mistakenly think they're the same, we think that object must have its length changed during motion. Special Relativity's length contraction is just a faulty measurement. It's an illusion. Nothing else."

"So speed can cause illusion?"

"Anything can cause illusion. You walk into a shallow pool and see your legs appear shorter. When you go spearfishing, the fish's actual position is not where you see it due to refraction. You take a photo of a running athlete; depending on your camera shutter speed, aperture, and ISO, parts of the athlete's image could be missing, blurry, overlapped, or dilated. You put two fish into a small aquarium, and depending on your viewing angle, you can see five fish or more. Those images are illusions and not true representations of reality. Comparison or measurement can only make sense within the same reference frame."

## Homework

"I have some homework for you," Newton said.

"Why?"

"Can you summarize everything we've discussed in just ten questions?"

"Questions for whom?"

"For all scientists who support Special Relativity."

"Why?"

"To test if they can pass Special Relativity."

"Who can test them?"

"Everyone."

# CHAPTER 14
# QUESTIONS FOR SCIENTISTS

*The important thing is not to stop questioning.*
*Curiosity has its own reason for existing.*
*ALBERT EINSTEIN*

*One of the really tough things is figuring out what questions to ask.*
*Once you figure out the question, then the answer is relatively easy.*
*ELON MUSK*

*Why have scientists accepted a theory which contains obvious errors*
*and lacks any genuine experimental support?*
*LOUIS ESSEN*

*Can you answer my questions, Professor?*
*JOURNALIST vs. SCIENTIST*

"It's wrong," Newton immediately said when I showed him a list a few days later.

"You haven't read it yet," I protested.

"This is not a 10-question list, but a book of questions. It has almost 1000 questions."

"No, it doesn't, only 100 questions. Maybe a bit more," I explained.

"100 or 1000 is the same - it's not 10. Go fix it."

And I went home. It was our shortest session.

# 1. If Einstein's postulate of absolute light speed is wrong, invalid, or senseless, would that be the end of Relativity?

1.1. What does Special Relativity's absolute light speed postulate actually say?

1.2. Does it say light speed is always measured as approx. 300,000 km/s (186,000 mi/s in a vacuum) regardless of the source or observers' motion?

1.3. Does it mean, according to Special Relativity, the relative speed of two light beams (starting from opposite ends of a distance) traveling toward each other must be 300,000 km/s and not 600,000 km/s?

1.4. Does it also mean, according to Special Relativity, the relative speed of two light beams (starting from the same location) traveling alongside in the same direction must be 300,000 km/s and not zero?

1.5. Is that result obtained from Relativity's velocity addition formula (Vac = Vab + Vbc/(1 + VabVbc/c²))?

1.6. According to Special Relativity (and its velocity addition formula), what's a light beam's speed relative to a traveler traveling at speed v of 0.0001c in the opposite direction?

1.7. What if v is 0.3c?

1.8. What if v is 0.8c?

1.9. What if v is c?

1.10. According to Special Relativity (and its velocity addition formula), what's a light beam's speed relative to a traveler traveling at speed v of 0.0001 c in the same direction?

1.11. What if v is 0.3c?

1.12. What if v is 0.8c?

1.13. What if v is c? (Tip: Calculate Lim f(v) when v → c)

1.14. What does it mean when we say two light beams travel alongside in the same direction? Does that mean they're always side by side, neck to neck, hand in hand, during the whole trip? Does that mean their relative speed to each other is zero? Then how can they possibly move toward (or away from) each other at a speed of 300,000 km/s at the same time, according to the second postulate?

1.15. If the second postulate is interpreted that way, would that mean $0 = 300,000$?

1.16. Is a point in Einstein's space 300,000 km wide, according to Special Relativity?

1.17. Can you hold someone's hand and at the same time travel 10 km/h toward or away from that person?

1.18. Can you stay in London and murder someone in Melbourne at the same time? Does that reasoning mean the end of alibis?

1.19. During a marathon, can an athlete claim he reaches the finish line first despite his being last? Can he say, according to him, that the finish line is the same as the start line? If so, why does he race?

1.20. Surely Special Relativity's absolute light speed postulate doesn't mean that, does it?

1.21. What if Einstein only meant that, once intercepted, light speed is always measured at 300,000 km/s, but its relative speed can be calculated as faster or slower depending on observers' movement? In short, the postulate is only a warning about our measurement limitation, which could lead to miscalculation if wrongly applied, isn't it?

1.22. If so, why would Einstein endorse that faulty measurement as a postulate in a physics theory?

1.23. If you ask 10 pro-Relativity scientists about the meaning of that second postulate, would you get the same answers?

1.24. If their answers are different, why? What does that tell you?

1.25. How should we change the postulate to avoid any misinterpretation without changing it?

1.26. If you write a computer program containing a line that the computer cannot read, interpret, run, or execute the way you want, whose fault is that? Too good a programmer or too stupid a computer?

1.27. Who is to blame – sellers, buyers, or the product?

1.28. Is Special Relativity's second postulate, as understood the way it's written, non-executable and contradictory to the definition of speed?

1.29. Can one thing be two places at one time?

1.30. Can light be in two places at one time?

1.31. If light can be in two places at one time, why does it have only one speed but not two? If so, what is the other speed that is different from 300,000 km/s?

1.32. Can light have a speed that is not its speed?

1.33. Similarly, what evidence does the Michelson-Morley test result provide? Evidence of absolute light speed or evidence of illusion (of absolute light speed)?

1.34. What is the worst thing that can happen if we calculate light speed relative to another light coming from the opposite direction as 600,000 km/s, using Newton's velocity addition formula, without Special Relativity's time dilation, length contraction, or space–time curvature?

## 2. If Einstein's postulate of absolute light speed is wrong, invalid, or senseless, would Special Relativity's time dilation and length contraction equations become invalid?

2.1. Is it true that both Special Relativity's time dilation and length contraction equations are deduced directly from the second postulate of absolute light speed?

2.2. Will they fall if that second postulate falls?

2.3. What are t and t' in Special Relativity's time dilation equation?

2.4. Are t and t' interchangeable in that equation due to inertial motion's symmetry?

2.5. Is it possible that the correct interpretation of $\Delta t$ (difference between t and t') is only a calculation difference during motion when $v \neq 0$, meaning when $v = 0$ at the end of the trip, $\Delta t = 0$, $t = t'$, there would be no real-time dilation?

2.6. What are L and L' in Special Relativity's length contraction equation?

2.7. Are they interchangeable due to inertial motion's symmetry?

2.8. Is there evidence of real length contraction due to inertial motion at the trip end (when v = 0)?

2.9. Time dilation and length contraction go hand-in-hand in Einstein's equations; if there is no evidence of real length contraction due to inertial motion, why is there evidence of real-time dilation?

2.10. If you're a mathematician, what is the correct way to interpret Special Relativity's absolute light speed postulate?

2.11. If you're a mathematician, how many different time dilation equations can you deduce from that postulate, given its many different interpretations?

2.12. Mathematicians can prove an equation is true according to math. But who can interpret, apply, and prove an equation is true in reality?

2.13. Let's call Einstein's original time dilation equation $\Delta t = f(v)$. Should we also have $\Delta t = g(v)$, $\Delta t = h(v)$, etc., simply because t is no longer an absolute time, but a relative speed like everything else?

2.14. When talking about speed, we often ask which speed, because everything has a different speed relative to something. No one would ask which time, since time is absolute and there is only one time. But if now time is relative and everyone has a different time, of which time are we talking, whose time is it, and relative to who?

2.15. If one thing has hundreds of small parts inside, each has a different speed (relative to something), should it also have hundreds of different clocks inside (relative to different observers)?

2.16. Regardless of whichever time dilation equations, none could be used to predict the actual single reading on any manmade physical clocks, because the equation knows nothing about those clocks, such as Rolex, Casio, sundial, pendulum, digital, atomic, etc. True or false?

2.17. Using Einstein's time dilation equation to predict a clock's actual reading is a huge misinterpretation, false assumption, and illegal application, isn't it?

2.18. According to Special Relativity, speed is the only cause of time dilation and length contraction. Is it possible that in the future, we might discover another factor that can cause time dilation and length contraction?

2.19. Can gravity cause time dilation?

2.20. Can heat cause time dilation?

2.21. Can pressure cause time dilation?

2.22. Can manually winding a clock cause time dilation?

2.23. If not, why not? Why can nothing else, apart from speed, cause time dilation and length contraction?

2.24. If yes, there is another factor (expressed in another equation) that can cause time dilation, what will be a clock's final reading?

2.25. Does it mean that Special Relativity's time dilation equation alone cannot be used to predict a clock's final reading, which is the sum of many unknown factors?

2.26. What can it predict, then?

2.27. Is it true that an equation (even if it's true mathematically) has nothing to do with how it would be interpreted or applied in reality?

## 3. What is measurement, why do we need it, and how is it done?

3.1. Can we survive this world without measurement?

3.2. Can we go shopping, sell items, trade stocks, make cars, build hospitals, or fly spaceships if we know nothing about measurement?

3.3. We don't need measurement for trivial things, but when things matter, we need measurement. True or false?

3.4. The purpose of measurement is to tell how different one thing is from another when plain words are insufficient. True or false?

3.5. What makes one thing different from another? Its measurements?

3.6. If two things are different, they must have at least one different measurement about something between them. True or false?

3.7. Measurement is the process of associating numbers with physical values or phenomena. All numbers are different; therefore, one thing can only be associated with one number, not two, because one thing cannot be two different things, and one number cannot be two different numbers, unless we're talking about different things. True or false?

3.8. Everything has only one measurement unless it's two different things. True or false?

3.9. Can science make a difference in life, if it cannot tell the same or difference?

3.10. Can science make a difference in life, if it cannot tell one thing from another?

3.11. Can science make a difference in life, if it cannot tell right from wrong, real from fake, and saving lives from killing people?

3.12. Can science make a difference in life, if it cannot measure how different one thing is to another?

3.13. If two things are so different for a life-and-death matter, yet their measurement fails to pick up any difference, it's a huge failure and a tragedy, right?

3.14. Don't mistake one thing for another. One thing has many parts inside, many relationships with others, and many characteristics (such as shape, color, dimension, position, volume, smell, texture, weight, mass, temperature, pressure, and so on); each is another thing with its own measurement. Correct measurement of the wrong thing is still faulty measurement of the right thing, right?

3.15. How can we ensure the right measurement for the right thing?

3.16. We need correct object identification process and correct measuring devices to know exactly what to measure and how to measure, right?

3.17. We measure by comparing different things with the same standard thing, right?

3.18. What is that standard thing called? Is it a ruler or measuring device?

3.19. Can we change different rulers with different measurement units when measuring?

3.20. Measurement only makes sense if we have correct rulers with the same units. It's not easy, since any physical things can deform, including rulers. But if rulers are deformed, they're no longer rulers, but faulty rulers, and cannot be used. Their measurement cannot be trusted. True or false?

3.21. Similarly, we cannot use corrupted judges to judge corruption. Any human can be corrupted, but judges can't. Corrupted judges have to be removed. Their judgment cannot be trusted. True or false?

3.22. In sports competitions, we need umpires. Any human can be biased, but umpires can't. Biased umpires are no longer umpires, and have to be removed. Their decisions cannot be trusted. True or false?

3.23. To measure length and speed, changes and deformations, same and difference, we need absolute space–time, absolute clocks, and absolute rulers. Space–time curvature, time dilation, and space contraction (or length contraction) are all signs of a corrupted measurement system and cannot be used. Their measurement and description of the world cannot be trusted. True or false?

## 4. What is faulty measurement, and how can it be stopped?

4.1. What is the difference between true measurement and faulty measurement?

4.2. Can faulty measurement kill?

4.3. The patient requests to remove his right kidney, and his surgeon removes his left one instead, saying direction is relative. What would you do if you were the patient?

4.4. A nurse wrongly gives a patient 500 mg injection instead of 5 mg prescribed by the doctor, causing the patient's death. Is that murder due to faulty measurement?

4.5. A plane crashed landing at 1000 km/h in extreme weather with no visibility due to a faulty cockpit speedometer which displayed 300 km/h. Is faulty measurement to blame for the disaster?

4.6. If a car travels toward a pedestrian at a speed of 100 km/h, but the pedestrian, due to his poor vision or wrong calculation, thinks the car's speed is only 10 km/h when he crosses the street, what would happen when he got hit?

4.7. True measurement doesn't always save lives, but it's still better than false measurement. True or false?

4.8. A drives a car toward B at a speed of 100 km/h, but B calculates A's speed as only 10 km/h. If B never crosses the street and never gets hit or hurt, is it okay for him to stand on the footpath thinking that the car's speed is 10 km/h?

4.9. If you only think and never act on your thinking, what's wrong to think you're right even if you're wrong?

4.10. How can we prove someone is wrong if no one takes action to get hurt or punished for wrong action not yet taken?

4.11. Thinking without action is always right, even if it's wrong. Right or wrong?

4.12. Without a reality check, all theories are right even if they're wrong. Right or wrong?

4.13. Can we apply the same to Special Relativity? While an object is moving far away and has nothing to do with us, we can measure its length and time in whatever way we want based on any imaginary system we create, and it's always true in that system because no one gets hurt, and nothing matters. Is that the essence of Relativity?

4.14. What would happen when that object hits us?

4.15. When nothing matters, everything is true, even if it's false. True or false?

4.16. Talking about things that don't matter, why do we need science?

4.17. Talking about things that don't matter, why do we need measurement?

4.18. Life is relative, but when do we need to be absolute?

4.19. Talking about things that matter, if faulty measurements can kill, do we want to stop that?

4.20. How can we stop faulty measurements?

4.21. How can we determine faulty measuring devices?

4.22. We don't always know if measuring devices are faulty, but sometimes we can. What are some examples of faulty measurements that anyone can tell immediately?

4.23. If you jump on a scale that says 1500 kg, would you know the scale is faulty?

4.24. If you walk with a friend in a park and a distance meter says your distance to your friend is 100 km, would you know it's a faulty device?

4.25. To avoid faulty measurement, first, we must realize that one thing can only have one true measurement and everything else is faulty. True or false?

4.26. If you weighed 100 kg and 200 kg on two different scales, one must be faulty. True or false?

4.27. If you weighed 100 kg on one scale and 100.0002 kg on the second scale, you wouldn't say it's faulty due to its small error margin. True or false?

4.28. If a table is measured as one meter and two meters long on two different rulers, one of them must be faulty, right?

4.29. If room temperature is measured as 20°C and 100°C on two different thermometers, one of them must be faulty, right?

4.30. If the same time-lapse is measured as one hour and three hours on two different clocks, one of them must be faulty, right?

4.31. A lawless society has no law. It's messy, chaotic, and violent. To fix it, we create a lawful society with a law allowing everyone to have his own law and take matters into his own hands. What difference can that law make? Or is it just a naming difference from a lawless to a lawful society?

4.32. Life without science is messy, chaotic, and confusing. It has no measurement, no accountability, and no certainty. But if we create a science that allows everything to have its own space, its own time, its own clock, its own measurement, and its own interpretation of the same and difference in the name of relativity, what difference can that science make?

4.33. Science is built on correct measurement; if measurement is not correct, what is left of science?

## 5. What is length, and how is it measured in space?

5.1. Can a length have two different lengths?

5.2. Can we use a curved ruler to measure a straight length which is measurable by a straight ruler?

5.3. Would a curved ruler show a different reading from a straight ruler?

5.4. If it does, does it mean one thing can have two different measurements or, simply put, a faulty measurement?

5.5. If it doesn't, why can't we use a straight ruler?

5.6. What is the purpose of creating a curved ruler?

5.7. If mass curves space–time, then how can we on Earth receive light straight from the universe? Why wouldn't light follow space–time curvature and bounce away from us?

5.8. If mass curves space–time, what space is occupied by Earth? Is Earth spaceless or out of space or nowhere? How can we address a location on Earth, when Earth is spaceless? (See Figure 13.2.)

5.9. What is the purpose of Einstein's space–time curvature, given space–time is an imaginary reference system to locate, navigate, and measure physical things in the universe?

5.10. Similarly, should we use curved longitudes and altitudes on Earth? (See Figure 13.6.)

5.11. Is Einstein's length contraction the same as space contraction?

5.12. Does space contraction here cause space dilation somewhere else? (See Figure 14.1.)

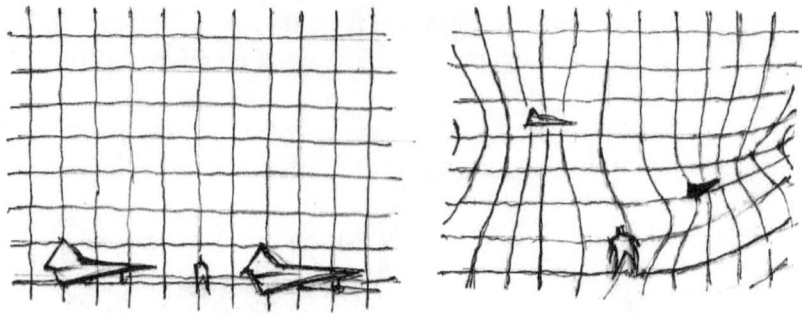

*Figure 14.1 Does space contraction here cause space dilation somewhere else?*

5.13. Can my space be contracted or dilated by other things moving around me, even when I'm not moving?

5.14. If not, why not? Where does space come from?

5.15. If yes, why does Special Relativity's length contraction equation have only one variable of my speed despite millions of things moving around me?

5.16. How can Special Relativity let me calculate my space contraction using only my speed while millions of things moving around me can also affect my space?

5.17. If we put a 20-meter length into a 60% contracted space (in the same direction), would that length now be contracted to 12 meters?

5.18. Why? Why does space contraction cause length contraction?

5.19. Despite its new length of 12 meters, if its contracted ruler still reads 20 meters for that length, is that reading a faulty measurement by definition?

5.20. Does Special Relativity allow everything to have a length and also a local space for a different measurement?

5.21. What's different between local space and global space?

5.22. What is space, and how is it different from length?

5.23. If ten high-speed spaceships fly at different speeds along a space highway, how is length contraction calculated on each spaceship, the highway, and the global space?

5.24. Is space–time curvature an excuse for faulty measurement?

5.25. Is it an illusion of reality?

5.26. Is space–time curvature a fancy term for a corrupted measurement system?

5.27. Is it true that sometimes we create great-looking but poorly defined concepts and end up having no idea what we're talking about?

5.28. A mechanic fits six-inch tires into an airplane's landing wheels against strict 20-inch specifications saying that his tires are actually 20 inches according to his space contraction equation. What would you do if you were his boss?

## 6. What is time, and how is it measured?

6.1. What's the purpose of creating a word called "time"?

6.2. Why do we need to measure time?

6.3. Everything changes in our world, and we need to measure their speed – rate of change – because that matters. Speed measurement is impossible without time measurement. True or false?

6.4. To measure time, we need correct timekeepers or clocks, right?

6.5. Clocks that run slow are faulty, aren't they?

6.6. Does that mean all length contraction and time dilation recorded on rulers or clocks for the same thing are signs of faulty devices and must not be used in scientific measurement?

6.7. What is Einstein's time dilation? Is it the same as clocks slowing or slow aging?

6.8. If time dilation is clock slowing, does that mean inertial motion can change a clock's speed against Special Relativity's first postulate?

6.9. A worker comes to work exactly one hour late every day, blaming his clock's time dilation during travel. What would you do if you were his boss?

## 7. What is speed, and how many speeds can you see in one thing?

7.1. Is (distance) speed equal distance over time?

7.2. Can we measure (distance) speed without knowing how to measure distance or how to measure time?

7.3. If distance has only one measurement, time has only one measurement, speed must have only one measurement. True or false?

7.4. Everything has only one measurement; therefore, everything has only one speed (unless it's two different things). True or false?

7.5. Distance speed is a relationship between two things, and never a standalone characteristic of one thing. True or false?

7.6. To make sense of speed, we must clearly state which speed of what is relative to what. True or false?

7.7. In general, apart from distance speed, we also have many different types of speed, such as the speed of growth, the speed of production, the speed of typing, the speed of climate change, and so on; therefore, speed can generally be defined as the rate of change. True or false?

7.8. Is speed measured by change over time?

7.9. How many speeds can you see from one thing?

7.10. How many changes can you see from one thing?

7.11. A is running on a gym treadmill at a speed of five km/h. B is sleeping on the gym floor next to that treadmill. What is A's speed relative to B? What's B to A? What's A's heart rate compared to B's despite their same location? (See Figure 14.2.)

7.12. If A is running on the treadmill, is he running or stationary, according to B?

7.13. If one thing cannot be two different things, why is A seen as running and not running?

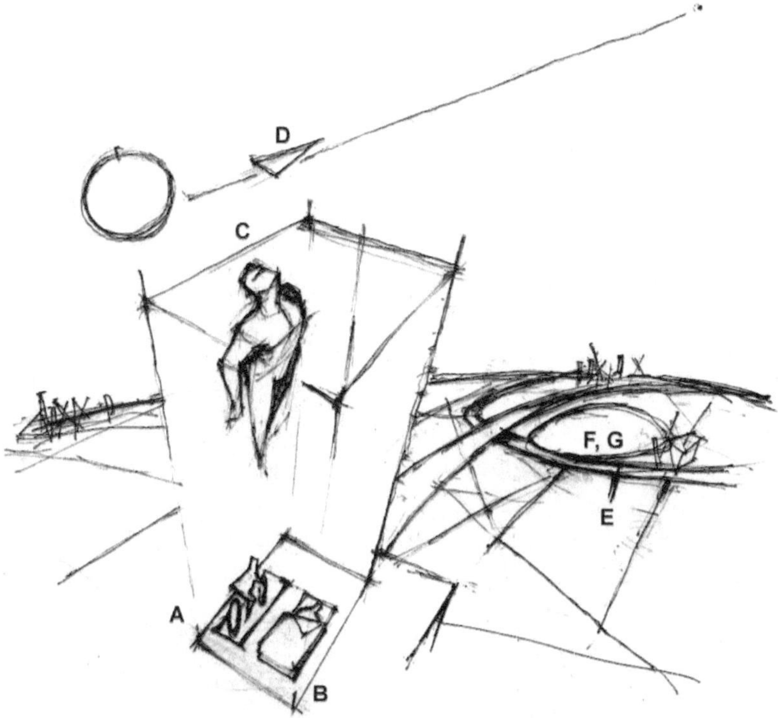

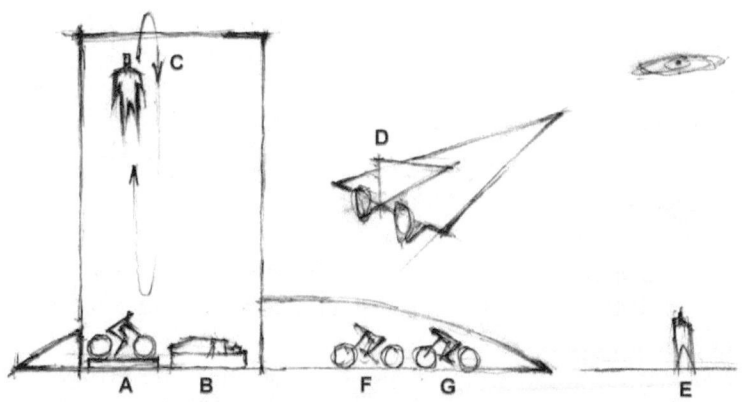

*Figure 14.2 How many speeds can you see?*

*Figure 14.3 What is C's vertical and horizontal speed relative to E?*

7.14. Is it reality that one thing can be two different things? Or is it just an illusion?

7.15. The gym is located on the ground floor of a 500-meter-tall skyscraper, C is a superman flying vertically inside a lift tunnel from the gym to the top floor at a speed of 100,000 km/s, back and forth, non-stop. What's the vertical speed of C relative to A and B? (See Figure 14.3.)

7.16. What's C's horizontal (or lateral) speed relative to A and B, when all stay inside the same building? Are they all zero?

7.17. D, a dead mouse, placed inside an interstellar spaceship, flies away from Earth to a Nebular galaxy at a speed of 200,000 km/s for 10 years. What is D's speed relative to A, B, and C? Are they all 200,000 km/s, given A, B, and C are all staying inside the same building on Earth?

7.18. E lives on a farm 100 km from the city building. What's C's vertical speed relative to E? What's C's actual speed relative to E, given the C–E distance is always 100 km from the city to the farm, regardless of how fast C is flying vertically inside the building lift?

7.19. Inside a velodrome nearby, F and G are racing their bikes, one after another, in the same direction, at the same speed 30 km/h on a single track. What's their speed relative to each other, presuming their distance is always two meters apart? Is it zero? Does that mean they're both perceived as stationary to each other?

7.20. After a few warm-up laps, F accelerates and moves up ahead of G at half-track length on the opposite side, then maintains the same speed of 30 km/h as G's. What's their speed relative to each other now, presuming their distance is always 100 meters apart? Is it still zero? Does that mean they're both perceived as stationary to each other?

7.21. A spectator, sitting outside that velodrome and looking through a trackside window, notices each time F goes past him. G also goes past in the opposite direction on the other side, and he measures their speed as 60 km/h relative to each other. From the top view, what's their relative speed now, presuming their distance is still always 100 meters apart? Zero, 30, or 60 km/h?

7.22. Do F and G travel in the same direction or opposite directions? (See Figure 14.4.)

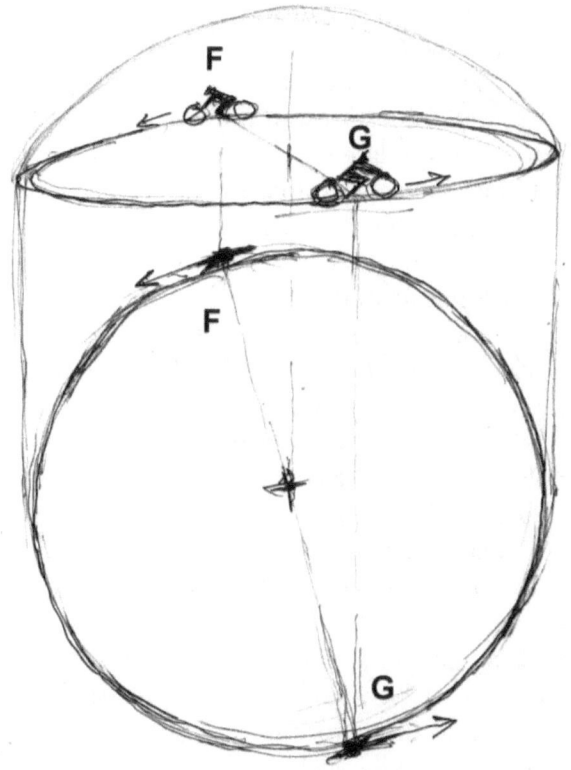

*Figure 14.4 Do F and G travel in the same direction?*

7.23. Is seeing always believing?

7.24. What does it tell you about speeds?

7.25. Can one speed tell the whole picture about objects' behavior?

7.26. If speed equals distance over time, does it always mean Vxy = Vyx?

7.27. Talking about distance speed between X and Y, how many speeds can you see?

7.28. Is it true that we might see many different speeds in one thing, not because one movement has many measurements, but because one thing has many images seen by different observers from different views, resulting in different speeds for different things?

7.29. One thing has only one measurement, but one thing can have many images with many different measurements. True or false?

7.30. Each image can reach different observers at different times, but does it mean time is relative? Or is image relative? Or is speed relative?

7.31. Using one thing's image measurement to presume the same for that thing's other measurements is a huge misinterpretation. True or false?

7.32. Is Special Relativity's second postulate (about absolute light speed) a wrong description of reality, resulting in faulty measurement?

7.33. Is Special Relativity's simultaneity relativity just visibility relativity due to different images observed by different observers and the lack of absolute time?

7.34. Let's go back to the story about A, B, C, D, F, and G. After 10 years, they all reunite back on Earth. We check their clocks and discover some faulty clocks displaying different times. What makes clocks faulty?

7.35. Can we use Special Relativity's time dilation equation to predict how faulty clocks will be? Or we can't because it's not an inertial motion? But if we forcibly apply the equation regardless, which speed should we use as input? Which speed should we input as moving? Which is stationary by definition? What is output as time dilation? How do we interpret that output? Would it mean a clock will have different values, vertically and horizontally? Would that only happen during the trip? Which clock is moving and which is not? Or is the only thing it can prove is that the speed of one second will have different values, vertically and horizontally, depending on different speeds obtained by different observers, to fill up the gap caused by the absolute light speed?

7.36. If the value of one second changes physically, does it mean faulty measurement?

7.37. How does faulty measurement affect human aging?

7.38. Would the dead mouse come back younger or older if its clock runs wild?

7.39. Does math know? How would math know if we humans don't know the meaning of an equation variable and its interpretation into reality?

7.40. In a concert, everyone would hear the same song played by the pianist regardless of their different seating. But light is different from sound. Everyone, depending on their respective seating, would see different images from the pianist. Some see his face. Some see his profile. Some see his back. Some see part of his face obscured by the piano. Light has many different faces. We can only see the face facing us, but never other faces. We can measure the light coming directly to us, but we cannot measure other lights from other faces that don't come to us unless we use a mirror to interfere with their paths. A moving object under light would produce images everywhere, so we can measure that object's movement by tracking its images. But a photon of light going past us doesn't produce another second photon sideways for us to see or to measure the first photon. We cannot see it; we cannot measure it. Don't confuse the light going past us with another light coming to us. They have different faces. They are two different lights. Their measurements are different. True or false?

7.41. Could we have mistakenly measured this light for another light (or this face for another face) in the Michelson-Morley experiment, resulting in faulty Special Relativity's absolute light speed postulate?

7.42. Can one thing travel at two different speeds relative to the same observer at any time?

7.43. If it can't, does it mean Special Relativity's second postulate is wrong, invalid, or meaningless?

7.44. If it can, how can we introduce speed limits?

7.45. If a traveling car's speed is measured as 10 km/h and 200 km/h on two different speedometers, one of them must be faulty, right?

7.46. Can we dodge the speeding fine for actually driving a car at 200 km/h relative to the ground and say our speed is only 10 km/h according to our calculation?

7.47. To reduce space traffic accidents, we enforce speed limits with heavy fines for speeding space travelers ($1 billion for speeding over 100,000 km/s). Is there a way to measure and record their high speed so no one can dispute it?

7.48. Is that why we need absolute measurement when things matter?

7.49. Is Special Relativity's second postulate contrary to the definition of speed?

7.50. If you're a scientist, can you design a test to prove that you can catch a light beam sooner if you're moving toward it than someone not moving?

## 8. What is time dilation?

8.1. According to Relativity, time slows down in motion as clock retardation. If people on a high-speed spaceship conduct tests to measure the light speed, vertically and horizontally, what is the result based on their slower clocks? Still 300,000 km/s, or something different – less or more?

8.2. If it's different, does it mean we can tell if we're traveling or not in inertial motion?

8.3. If distance speed between two persons cannot determine their heart rate, blood pressure, oxygen level, or any health data, how can it determine their aging?

8.4. Would you look younger by just altering your age on your birth certificate?

8.5. You might look younger by taking magic tablets to slow your aging process, coloring your hair, eating healthy food, or having cosmetic surgery, but how does that slow your clocks?

8.6. Is it possible Einstein's time dilation is not clock slowing?

8.7. If time dilation is not clock slowing, what can it show?

8.8. If time dilation is clock slowing, does that mean faulty clocks with faulty time?

8.9. Should we use faulty devices in any scientific measurement?

8.10. Why would we use faulty clocks to measure time in science?

8.11. Is all evidence of Einstein's time dilation in fact evidence of faulty clocks?

8.12. When a man dies, his clock would not necessarily die; then why would a slowing clock affect human aging?

8.13. Do we have any equation showing that time dilation equals human aging?

8.14. How can a math equation predict a clock's reading when math knows nothing about that clock, its brand, mechanism, structure, design, power source, or working requirements?

8.15. How can a math equation predict someone's current age when math knows nothing about that person's DOB, health, lifestyle, or diet, and cannot even tell whether that person is now alive or dead?

8.16. How old is a man if he was born 50 years ago?

8.17. How old does a man look now if he was born 50 years ago and died 10 years ago?

8.18. Is it possible that a 50-year-old man might look 40 or 60, depending on many life factors?

8.19. If physics cannot make any prediction on the appearance of someone's face despite his real age, how can a math equation?

8.20. Does time dilation violate Special Relativity's first postulate?

8.21. Except for time dilation, does Newton's world still allow other speed changes, such as faulty clocks running slower or faster, human aging slower or faster, human appearance looking younger or older, human longevity slower or faster, etc?

## 9. How long does it take for two light beams A & B meet from a three million km distance, if both travel toward each other at a speed of 300,000 km/s relative to the ground (using Newton's measurement units)?

9.1. At a speed of 300,000 km/s, is it true that light beam A has traveled 1.5 million km forward in five seconds, according to the definition of speed?

9.2. At a speed of 300,000 km/s, is it true that light B has traveled 1.5 million km backward in five seconds?

9.3. Is it true that light A and B would meet in five seconds?

9.4. From a distance of three million km, if light A and B meet in five seconds, does that mean A's speed relative to B is 600,000 km/s, according to the definition of speed?

9.5. If someone calculates as 300,000 km/s in Newton's space–time, is that a faulty calculation?

9.6. If a speedometer detects that speed as only 300,000 km/s, is that a faulty measuring device?

## 10. What is the question?

10.1. A man is exactly one foot taller than everyone in the world. What is his height?

10.2. A city is exactly 10 km away from all cities in the world. Where is it?

10.3. Something is exactly 100 kg heavier than everything in the world. What is its weight?

10.4. Everyone with different heights is measured the same as one meter high according to a ruler. Who has a different height?

10.5. A dictator with a very low IQ asks a mathematician to design an IQ test such that no one can truly score higher than the dictator. Who is smarter?

10.6. Is it possible to find a great actor to make nonsense sound so great, like something beyond human understanding?

10.7. Should scientists and users learn to understand their language first, before using language to sell or buy discoveries?

10.8. If zero division is allowed, can we prove that 12 = 300?

10.9. If number dilation is allowed, can we prove the sum of any numbers is always less than 300,000?

10.10. If time dilation and length contraction are allowed, can we prove something can always fly at 300,000 km/s relative to everything, regardless of how fast they're moving toward or away from it?

10.11. A king, while drunk, makes a revolutionary law allowing all drivers to go through traffic lights regardless of green or red, as long as no accidents happen. Soon after, a man drives through a red light, kills someone, and gets a 10-year jail sentence because the judge says he misinterpreted the law. Another day, a woman drives through a green light, kills someone, and gets a 10-year jail sentence because the judge says she misinterpreted the law. After that, everyone becomes scared and confused, so they decide to drive like before, with green to go, and red to stop. Since then, no more accidents. Now everyone is so happy, thanks to the "correct" understanding, "correct" interpretation, and "correct" application of the king's traffic light revolution. What's wrong with a "wrong" revolution, given its "correct" interpretation and application?

10.12. Is Special Relativity the same revolution?

# CHAPTER 15
# NEWTON'S FURY

> *When you come out of the storm,*
> *you won't be the same person who walked in.*
> *That's what this storm is all about.*
> *HARUKI MURAKAMI*

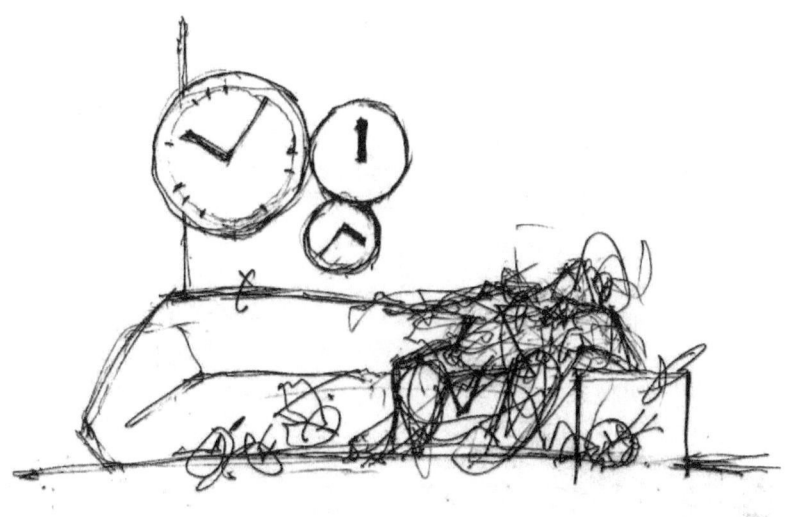

*Figure 15.1 Madman in a crazy world*

Nothing lasts forever. Something finishes, something else begins. We'd been talking for several weeks now, and I guess time was running out. I came late one rainy day, rushing from work, and forgot to get his meal. I said sorry and promised to bring it later. Not only that, I hadn't revised the list. Looking at the rain, I couldn't tell the difference. It was messy. Questions were lengthy. He was not happy.

"Can you summarize Special Relativity in 10 words?" asked Newton.

"You did."

"Can you solve it in 10 words?"

"Impossible," I said.

He looked disappointed.

"I had a dream last night," I casually said, changing the subject.

"What dream?"

"A trial against Special Relativity."

"You mean a public inquisition?"

"Kind of. It was out of this world, far from this time, like the story you told that was set in 3001."

"Another millennium." Newton chuckled sarcastically.

"Just a dream, anyway. It was brought up by a group of students accusing all universities of teaching fallacy and suppressing free speech."

"Sounds right. Tell me more."

"Many professional physicists were called to give evidence, together with many witnesses from the public. The trial lasted for several weeks. Many questions were raised…"

"What questions?"

"Every single question we've discussed here."

"What was the outcome?"

"Don't you want to know what led to the outcome?"

"I know all the arguments. No need to repeat. Just tell me the court's judgment."

"But it's only a dream."

"What's the difference?"

I hesitated.

"The chief judge criticized lots of evidence given by the professional physicists. He said their evidence was unclear, inconsistent, and even self-contradictory. He also considered all the evidence about censorship and the abuse those students had suffered for voicing their criticism against

Special Relativity. But after all, the judge was reluctant to convict Special Relativity as a fallacy. He blamed professionals' inadequate understanding of the theory, their lack of precise description of the universe, and their poor communication with students. He dismissed the case. He said that Special Relativity can still be true theoretically..."

"That is bullshit," Newton shouted. He violently kicked his table and threw everything off.

I froze in shock, eyes wide open at his tirade. He kept screaming.

"That is a wrong dream with a wrong outcome. What's wrong with all your scientists from your stupid world that allow this stupid joke to drag on for so long? Why has no one got the guts to question where this absolute light speed postulate came from? Does it come down from heaven? Does it rise up from hell? What does it mean? Why does no one in your world realize that the Hafele-Keating test is nothing but data manipulation? How much longer will the public let this theory fool them? If Einstein's time dilation equation was correct, why wouldn't you use it to fix time? Why would you want faulty clocks with time dilation instead of correct clocks without time dilation? Why would you want faulty rulers with length contraction instead of correct rulers without length contraction? Why does no one in your world realize math has been misused to wrongly validate a wrong theory built from a wrong postulate? Can't you see all equations are symmetric in inertial motions? How can symmetry produce any difference? It's cheating. It's rubbish. Math has nothing to do with reality – it's a cover-up to stop people from questioning the stupidity of absolute light speed.

"Did you know that speed is a relationship? Did you know that nothing has a speed without reference to something else? And what about simultaneity? The question is not whether it's relative or absolute, but what does it mean? If simultaneity is defined as events happening at the same time, then how can you tell if events are simultaneous or not simultaneous without an absolute same time system? How can you talk about simultaneity if you can't define what time is? If each event has thousands of mini-events inside, what events are you talking about?

"What about space curvature? What a load of garbage. Where does space curve from and into? Can you measure its curvature? If you can, does that mean you have to use Newton's flat space to measure any curvature against it? If so, why don't you just use Newton's flat space to say a straight is a straight and a curve is a curve? Why would you use Einstein's so a straight line is now a curve, and a curve is now a straight line? What the hell is that? Another fourth dimension of language?

"Does using fancy words make this world a better place? Does it make you understand better or only make you sound superior, so you can fool more people about the same world but different names? Can you change the world by using fake language? It's a crime to teach students to falsify measurements. It's wrong. It's fraud. It's all bullshit. You hear me? You *** hear me?"

"I'm sorry. I'm sorry. Please, Mr. Newton, please calm down."

"I don't need you. I'm sick of all your stupid questions. I don't need any more idiots from your world. I've had enough. Get your *** out of here. GET OUT!" Newton screamed.

I panicked at his fury and realized leaving was the only choice.

"Yes, yes, I'm leaving now. It's okay. Please calm down, Mr. Newton. Please."

***

One hour later, when the rain had stopped, I returned – just to make sure he was okay and to bring his meal, which I had promised. The door was not locked. From outside, I could hear his snoring. I gently stepped in and placed his meal on

176

the kitchen table. He had fallen asleep on his couch. I tried to tidy up his room by gently picking up broken glasses and damaged clocks from the floor. The room was messy, dirty, and smelly. So were his clothes and the wig he always wore.

He is not Newton. His name is Isaac Newton. But he is not the Newton that we know, just a mental patient suffering from schizophrenia I had met many years earlier. He used to be a math teacher in junior high school but retired long ago due to his illness.

After long rehabilitation at St Albans Continuing Care Unit, he had transferred here to this tiny granny flat and lived alone, with no relatives and no friends. Often missing his meds and relapsing quickly, he'd been put under a psychiatric treatment order and was frequently visited by staff from a local community mental health clinic. Troubled with delusions, he often disguised himself as the great Isaac Newton, talking about physics stuff no one can understand. He was also on an administration order to have his finances managed by a state trustee because he could not control his disability support pension, and would often spend everything on clocks.

I looked at him and realized my time was up. So I took a piece of paper and wrote a brief note: "Thank you, Sir, for your time. I'm sorry for upsetting you with my wrong dream. I hope reality will be a better one – and correct one. Anyway, I won't disturb you again. I hope you know that I'm sincerely grateful for your lectures and the answers you've helped me to see. Thank you."

I stood up and walked to the door. He was so peaceful now in his sleep. I took a last look and couldn't help wondering. Don't you feel like a madman sometimes when sane people cannot make sense of your world? (See Figure 15.1.)

# CHAPTER 16
# TRIAL OF THE CENTURY

*Our lives begin to end the day we become silent*
*about things that matter.*
*MARTIN LUTHER KING*

*If you thought that science was certain – well,*
*that is just an error on your part.*
*RICHARD P. FEYNMAN*

*Do you teach fallacy and suppress freedom of speech, sir?*
*BARRISTER vs. PHYSICIST*

The trial against Special Relativity finally happened. A group
of 21 students had taken legal action against many of the
world's top universities for teaching fallacy and suppressing
freedom of speech at the International Court of Scientific
Integrity, Netherlands. (See Figure 16.1.)

The trial lasted several weeks, with many top scientists,
authors, and professors subpoenaed to provide evidence
about Special Relativity and its interpretation. Both sides
tried to convince the public. Questions asked, answers given,
and disturbing stories heard.

On the last day, Ms. Viazovska, the barrister for 21
applicants, delivered her closing address to members of the
Court.

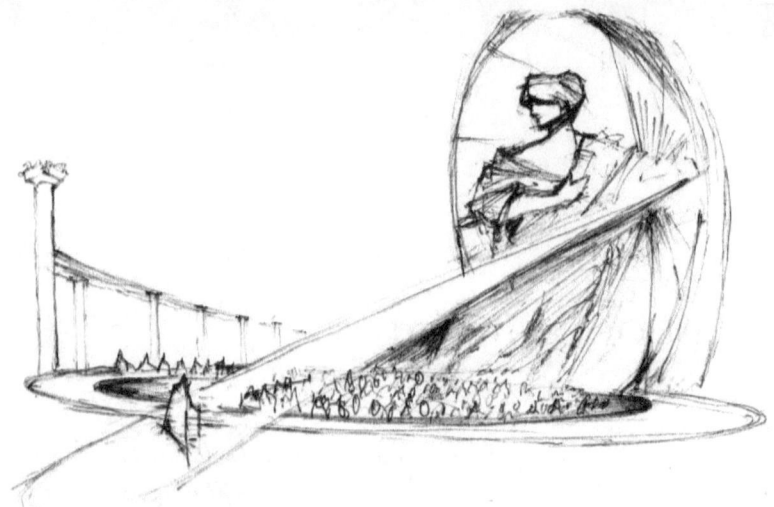

*Figure 16.1 Trial of the century*

## Why are we here?

Ladies and gentlemen,

This trial is not an attack on Einstein or the physics establishment.

This is actually to honor Einstein's legacy, to follow in his footsteps, to question our knowledge, and to promote a true science that can make a real difference for all of us.

Einstein is a genius. We all know that, and we all love him. His contributions to science, mankind, and social justice are gigantic. He's an aspiring role model for many generations of scientists who dedicate their lives to further human knowledge, world peace, and humanity.

Among thousands of statements he made in his lifetime, if there were a couple of mistakes, they would not taint his image and reputation.

After all, Einstein was also a human. Any human can make a mistake.

But if a simple mistake repeated for so long has become the biggest joke in physics, it needs to be stated, stopped, and corrected so we all can move on and focus on Einstein's other greater achievements.

This trial is about one single mistake – Einstein's misconception of time.

It was brought up by a group of 21 students who love Einstein, science, and truth.

They thought those were identical until the day they hit Special Relativity and failed.

Many have quit physics. Some abandoned their dreams. Some have moved to other disciplines. One student had his PhD thesis threatened unless he stopped criticizing Einstein's Special Relativity.

Nothing is unusual about students who take on hard subjects and fail. Someone fails at something.

But there's a difference between failing what is too hard and refuting what is nonsense.

This case is about the latter.

Our argument is those students have not failed, but our education has.

We have taught them a wrong theory – full of contradiction, misconception, and misinformation.

After all, Special Relativity is not the hardest subject to study. There are many harder subjects with greater impacts on our lives, such as medicine, biochemistry, math, engineering, psychiatry, psychology, programming, finance, sociology, law, economics, politics, and AI.

But Special Relativity is the hardest to accept because behind its mysterious, intriguing, and fascinating cover, it's dead wrong – it's a fallacy.

Let me summarize all the evidence in this case to help you reach your judgment.

## 1. Theory of Confusion

Science is to make a difference. But the only difference Special Relativity has made is confusion due to its ambiguous language.

Special Relativity claims that inertial motion can cause time dilation and length contraction.

181

But what is time dilation? What is length contraction?

To us, time dilation means youthful looks, a healthy body, and extended longevity.

To us, length contraction means, simply put, a 10-meter-long stick yesterday is shortened to nine meters today, based on the same measurement system and same reference system.

But does Einstein mean that?

Thirty-six pro-Einstein professional physicists have given evidence in this trial to support Special Relativity. Many are theoretical physicists, well-known authors, and celebrities. They all praised Special Relativity as a revolutionary theory in modern physics with a mountain of proven evidence.

However, when asked such basic questions during cross-examination, their answers are all different and even contradictory.

Why?

A simple question, such as whether time dilation means clock slowing, some say yes, some say no. For those who say yes, when asked which clocks, what brands, where, and how those clocks are created, they struggled to explain. When asked if we know exactly what makes clocks run slow, then why can't we use that knowledge to stop clocks from running slow, they cannot answer.

One mathematician even says Einstein's time dilation cannot be shown on manmade clocks, but is a purely abstract concept for calculation that has nothing to do with reality.

When asked if the same happens to length contraction, he says yes, while admitting physics is not his expertise.

And some physicists, who are not mathematicians, can use math to prove physics.

Some declare length contraction is only a measurement contraction due to different reference frames but no physical change in reality.

182

Can you believe all that? Do we speak the same language? What version of the English language are they using?

If mainstream physicists cannot fully understand the theory's language, interpretation, and prediction, how can they agree on the evidence to prove that theory?

The problem is not about those who don't understand the theory, but those who claim to understand still give contradictory interpretations of the theory.

Truth is, unlike any scientific theory in mankind's history, Special Relativity is the first and only one that deliberately alters the meaning of some basic concepts to make a faulty impression about a new world.

It's a theory of confusion.

## 2. Theory of Fake Evidence

Special Relativity claims inertial motion can cause time dilation and length contraction, but the truth is that no physical evidence has ever confirmed that.

No evidence has ever confirmed length contraction due to inertial motion.

We have evidence of objects' length contracted, expanded, and deformed due to extreme heat, force, and pressure, but none due to inertial motion.

For over 100 years, no professional physicists have ever mentioned evidence of length contraction due to inertial motion.

Length contraction and time dilation go hand in hand in Special Relativity. If length contraction has no evidence, why should time dilation?

Truth is, there is only fake evidence.

We have evidence of faulty clocks (such as clock retardation) under extreme conditions due to extreme force, gravity, and acceleration, but none due to inertial motion.

Time dilation doesn't exist, only faulty clocks.

Don't use a gun to shoot a cat dead and claim the sound has killed it.

Don't prove this theory by using evidence for other theories.

Correlation is not causation.

It's 100 years of fake evidence.

## 3. Theory of False Advertisement

Special Relativity is not the solution to youthful looks, a healthy body, and extended longevity, as falsely advertised and misled to the public.

Even if its time dilation is interpreted as clock slowing, it has nothing to do with human aging.

If a man has a heart attack and dies, his watch will not stop running. No evidence ever confirms a correlation between a slowing clock and human aging.

As a scientific theory, Special Relativity has produced only two math equations for time dilation and length contraction: ($t = t' \gamma$) and ($L = L'/\gamma$).

As far as math goes, the theory stops there.

Beyond that, how to apply those equations to reality is everyone's agenda.

It's no longer math, logic, or science. It's all about human business, imagination, and interpretation.

Truth is, Special Relativity's equations have no variables representing youth, health, and longevity.

$t$ does not represent youth, health, or longevity.

$L$ does not represent a real physical contraction; it is just a double measurement of the same length due to different reference frames.

They're all products of false advertisements.

## 4. Theory of Self-Contradiction

Special Relativity's two equations (time dilation and length contraction) are products of self-contradiction.

Special Relativity's first postulate states that the laws of physics remain the same in inertial motion – meaning inertial motion, and inertial motion alone – cannot change any physical object's behavior.

And yet, Special Relativity claims inertial motion can change an object's speed (clock speed) and length.

Special Relativity is a theory of self-contradiction for saying one thing and immediately saying the opposite.

In science, anything self-contradictory is nonsense.

## 5. Theory of Double Measurement

Science builds on correct measurement – because faulty measurement can kill.

Yet, Special Relativity is the first and only scientific theory that legalizes double measurement in the name of relativity.

According to Special Relativity, a 100-meter-long spaceship flying in the sky would have its length contracted to 90 meters, and its pilot would not know because his ruler (also contracted) still reads 100 meters. By definition, it's a faulty statement or faulty measurement.

In science, if there is a physical change but the measurement fails to detect, then by definition, it's a faulty measurement.

In science, if a measurement is correct when displaying no change, then there must be no physical change.

In science, we cannot say, "Yes, there is change," but measurement says, "No, there is no change," and both statements are correct.

Double measurement – a critical concept in Special Relativity – is faulty measurement.

A scientific theory based on faulty measurement is a faulty theory.

## 6. Misunderstanding of Relativity

Einstein was confused between relative and absolute.

Life is relative, fuzzy, and confusing. When it doesn't matter, we can live with it. At best, that relativity creates mystery, excitement, unpredictability, surprise, variety, and fun. It gives us different views. But at worst, it can bring chaos, disaster, and death. That's when and why we need science to intervene, to explain why things are relative, how to fix that, to make relative absolute.

We cannot stop "relative" by staying "relative." We need an "absolute" stance to stop "relative." Of course, what we find absolute today can be relative again tomorrow. And the same process repeats – the search for perfection goes on. But science must never deliberately go backward – to make absolute relative. It's never science's purpose. That's exactly Einstein's error – he's made absolute relative.

Einstein thinks that since life is relative, all scientific measurements (including correct ones) are all relative, and science should embrace them all (even faulty measurements). Einstein mistakes problems for solutions, obstacles for purposes, and illusions for reality. He accidentally uses Relativity as an excuse for faulty measurement, which can be fatal.

Luckily, faulty measurements in Special Relativity have yet to kill anyone, simply because it might take thousands of years before humans can produce spaceships traveling at close to light speed. Until then, Special Relativity has fooled everyone, even scientists and millions of innocent students, with the fake world of Relativity.

## 7. Theory of the Wrong Postulate – Absolute Light Speed

Special Relativity, based on a wrong postulate, produces wrong equations. Its second postulate of absolute light speed regardless of observers' motion as formulated in its velocity addition formula is just sheer nonsense.

The statement is meaningless.

Speed requires two points. We cannot say something has a speed without reference to something else. Nothing can have an absolute speed independent of observers' movement unless that thing can be at two different places at the same time. But if one thing can be in two different

186

places, that means two different things with two different speeds, not one thing.

By accepting that postulate, Einstein has accidentally committed the most basic error in all scientific knowledge – contradiction. One is not two. Same is not different. Law cannot be lawless.

We cannot say light has a speed that is not its speed.

No evidence can ever make sense of this light speed constancy against observers' motion.

The Michelson-Morley experiment is not proof of reality but an illusion of reality. Special Relativity, based on that illusion postulate, is a wrong theory.

## 8. Theory of Math Abuse

$1 = 2$ is false.

$1 = 2$ can be made to appear true – only in another different number system where those symbols have different values.

We can use either system, but not both. We cannot say $1 = 2$ (in another system), then say $1 \neq 2$ later (by using our system).

Mathematically, Special Relativity can be summarized as:
$$(c + v = c - v = c) \rightarrow (t = t'\gamma) \ \& \ (L = L'/\gamma)$$
where: $\gamma = \sqrt{(1 - v^2/c^2)}$

The first expression is wrong, based on Newton's absolute space–time system where "km" and "second" have absolute values. In Newton's world, everything has only one location and one speed – meaning one measurement. Zero is different from 1. One km is different from two km. One second is different from two seconds. That system matches our reality, where life is different from death.

However, that first expression can be made to appear true, based on some weird Einstein's system where "km" and "second" have relative values; that is, one km here is different from one km there, one second here is different from one second there. That Einstein's system has yet to be defined. Even if it had, Einstein cannot say that light has a speed which is c = 300,000 km/s because that value was obtained from Newton's system using Newton's language where "c," "km," and "s" have absolute values and light has only one single speed.

By concurrently using two different reference frames, two different measurement systems, and two different languages in the name of relativity, Special Relativity is nothing but math abuse.

Using math without rules is the wrong use of math.

It can kill.

## 9. Theory of Misconception – Time

If there is one word, and only one, that has made Special Relativity and will break it, it's time.

Einstein has misinterpreted time.

Time, in physics, is not a physical motion pre-existent in the universe that we can stop, manipulate, or change. Time, in physics, is an imaginary concept that constantly flows and cannot change because we deliberately imagine it that way, so that any change can be measured against it.

To measure change, we need a reference that never changes.

To measure difference, we need a reference that's always the same.

Time is created for that purpose – as a reference to measure speed. Therefore, time must be absolute, by definition.

If we want to change the world, we have to change its measurements and its speed, not time.

A boy has grown from one meter to 1.8 meters tall. That's only a real difference based on the same rulers with the same units.

A worker who used to work 12 hours a day now works eight hours. That's only a real difference based on the same clocks with the same rate.

A tennis player has increased his serve from 100 km/h to 200 km/h. That's only a real difference based on the same rulers and the same clocks.

The truth is Einstein confused time with speed.

Science is meant to make a difference.

Difference happens when its measurement changes – if it's a correct measurement.

Correct measurement is only possible by using correct measuring devices. But measuring devices, like anything physical, can change.

Clocks are manmade. Clocks can break down, turn faulty, and give wrong readings tomorrow, regardless of how well we design them today. As soon as we detect faulty clocks (or faulty rulers, or any faulty measuring devices), we must fix them, for faulty measurement can kill.

There is no time dilation, only faulty clocks.

Einstein mistook time dilation with faulty clocks.

Based on time dilation, Special Relativity is a theory of misconception.

## Conclusion

Ladies and gentlemen,

We cannot erase poverty on Earth by printing more money. We cannot use cars' outside mirrors to make objects appear further to claim length contraction. We cannot take a photo of a group of friends and claim those standing on the outside are bigger than the ones on the inside. We cannot use a photo of the Eiffel Tower standing on your palm to claim a revolution. We cannot wear different sunglasses to see the world in different colors and claim the world has changed. We cannot use a warped mirror to have a distorted view of our world and then claim space–time curvature. We cannot claim human longevity by using faulty clocks as evidence of time dilation.

You see? Seeing is not always believing. Images can be illusions. And life is full of illusions. Some are harmless. Some are fun. And some are dangerous – they can kill.

That's why we need science to tell us right from wrong, true from fake, and reality from illusion.

That's what Special Relativity fails – it has fallen victim to another illusion.

Instead of telling illusion from reality, Einstein sees illusion as reality.

The whole theory, built on the illusion of absolute light speed, creates more illusions about time dilation and length contraction.

There is no evidence of time dilation, but only evidence of faulty clocks. There is no evidence of faulty clocks due to inertial motion, but only to force, acceleration, gravity, heat, vibration, and so on.

Time dilation is not slow aging as advertised and misled to the public, but just another different measurement based on slower clocks – meaning faulty measurement.

Length contraction is not real; it is just a different measurement due to different reference frames when things travel far away and have nothing to do with us.

No evidence supports Special Relativity's other claims about space–time curvature, simultaneity relativity, and absolute light speed because those concepts contradict their own definitions. Self-contradictory concepts cannot have evidence.

Nevertheless, we have enough evidence of mainstream physicists' failure to understand the theory due to its ambiguous language, their misinterpretation of equation variables into reality, and their ignorance of questions raised by their critical colleagues, dissidents, whistleblowers, the public and students for over 100 years.

Special Relativity is a fallacy.

Teaching a fallacy is not evil, but failing students for questioning it is.

As one of the greatest scientists of all time, we strongly believe it can never be Einstein's true purpose and intention.

It's just an error on his part.

*Figure 16.2 Special Relativity: a perfect fantasy*

Ladies and gentlemen,

Behind these 21 students I represent today, may I remind you of thousands of anonymous scientists whose criticisms against Special Relativity have always been disregarded, censored, or ridiculed.

One of them is Louis Essen, an English physicist and the inventor of atomic clocks, who gave this humble, simple, and most accurate summary about Relativity:

> *I concluded that the theory is not a theory at all, but simply a number of contradictory assumptions together with actual mistakes.*

Let's not forget them.

Special Relativity is not true science. It's the greatest illusion. It's a perfect fantasy. (See Figure 16.2.)

# CHAPTER 17
# WHEN YOU WERE SEVENTEEN

*Promise me, you'll always remember:*
*you're braver than you believe,*
*stronger than you seem*
*and smarter than you think.*
*WINNIE THE POOH*

*Figure 17.1 The schoolboy in every one of us*

Win or lose, time will pass, and people will forget. But for those students attending that trial, if there is one thing they will always remember, it's the struggle of their youngest friend, who was last cross-examined by the defense barrister. The poor boy, regarded as the weakest link in the group, was quite skinny with long hair, had a nasty scar across his neck, and talked with an Australian accent.

"How do I say your name?" asked the defense barrister, Mr. McNamara.

"Tuan."

"How old are you?"

"Seventeen."

"Where do you come from?"

"Vietnam."

"You must be the smartest kid in your class?"

"No. I'm just average."

"What's your IQ?"

"I don't know."

"But if you sit for an IQ test, what score do you think you could get?"

"Maybe average."

"What school did you go to?"

"Nguyen An Ninh High School."

"Never heard of that. It must be the best high school in Vietnam?"

"No. It's just an average public school."

"I see. So you're an average student with an average IQ from an average school in an average country, and immediately you failed first-year science at an Australian university. Am I correct?"

"No, I didn't fail. I transferred to another course."

"Because physics is too hard for you. Am I right?"

"Many things in life are hard for me; I'm not lying. But I quit physics not because it's hard, but because I disagree. Special Relativity doesn't make sense to me. I'm here to study, but I cannot study nonsense."

"Do you need money?"

"Who doesn't?"

"I'm asking you."

"That is my answer. My question is my answer."

"Let me remind you, young man. I'm cross-examining your evidence. It's your chance to convince this court by explaining your evidence, and the only way to do that in support of your friends' case is by answering, not questioning. Am I clear?"

"Yes."

"Do you need money?"

"Yes."

"You're new in Australia. How long have you been there?"

"Six months."

"Are you working?"

"Yes."

"What are you doing?"

"I work two jobs, waiter and cleaner, apart from studying."

"What kind of cleaner?"

"May I skip this question, sir?"

"Why? Are you embarrassed by what you're doing?"

"No."

"Then what type of cleaner are you doing?"

"Toilet cleaner."

"I see. Looks like you're desperate. I bet you need money for many things, like rent, car, fridge, washing machine, and furniture, don't you?"

"Yeah, but that's not my priority."

"What is?"

"I have many poor friends in Vietnam. Many kids in rural North Vietnam's boarding schools don't have hot water. Some as young as eight have cold showers during freezing winter when the temperature can drop to 7°C. Kids skip showers because it's too cold. A hot water system there costs only $3,000 for a whole school with 300 kids. We've already helped a few schools. But we need..."

"That's enough. This is a legal proceeding about a scientific matter, not for charity fundraising. Can you please stop distracting the court?"

"Sorry, sir. I'm just..."

"Do you need fame?"

"No."

"You said you need money to buy a car, furniture, or help your friends. And fame can certainly bring money to help your friends. So, do you need fame?"

"Well, I've never thought of that."

"I put it to you; that is why you and your friends are doing all this – all about money, publicity, and nothing to do with Relativity, Einstein, or science. Is that true?"

"No."

"What is the real reason?"

"We do this because, first, we believe Special Relativity is wrong, and second, we cannot criticize Special Relativity in our universities. Money or loss, fame or shame, win or fail, they're not in our equation and certainly not our purpose."

"Have you ever heard of Bill Gates, Elon Musk, Mark Zuckerberg, Sergey Brin, Steve Jobs, Aaron Swartz, Jack Ma, and many more entrepreneurs?"

"Yes."

"Ignoring their fame and wealth, do you think they're very smart people?"

"Yes."

"Do you think they're much smarter than you?"

"Yes, of course."

"If Einstein was wrong, why did they say nothing?"

"You'd better ask them."

"I'm asking you. You said there are so many great and smart people out there who know about Einstein and his Relativity; if Einstein was wrong about time, why do you think those great people say nothing?"

"They must be too busy making money."

"C'mon. Don't you have anything better to say about those smart people?"

"That's what I meant. They must be busy doing what matters, like making a difference, creating jobs, changing life, instead of fixing time."

"Have you ever heard of Richard Feynman?"

"Yes, he's a great theoretical physicist and mathematician."

"What about Niels Bohr, Max Planck, Werner Heisenberg, Erwin Schrodinger, Paul Dirac, Enrico Fermi, Wolfgang Pauli, Hideki Yukawa, Max Born, Tsung-Dao Lee, Pavel Cherenkov, Lev Landau, Steven Weinberg, Peter Higgs, Kip Thorne, Terence Tao, Roger Penrose? Do you know who they are?"

"I only know some. Many are famous physicists."

"Yes. They're all Nobel laureates in physics, like Richard Feynman."

"They must be."

"And Stephen Hawking. You know who he is?"

"Yes. One of the greatest minds in modern physics."

"So you know they're all great scientists, physicists, mathematicians, right?"

"Yes, they're all my heroes."

"Heroes? Really? Why?"

"I love scientists. I admire them. I always respect great people with great minds who make a difference in life by erasing poverty, war, and violence on Earth."

"Do you always understand their works?"

"No."

"Why?

"Their works are too high for me."

"Well, you're a good boy, aren't you? At least you know your modest level and that you need to study more to understand their works, right?"

"Yes."

"If Einstein was wrong about Special Relativity, why do those great minds in physics say nothing?"

"You'd better ask them."

"I'm asking you. What do you think is the reason all those Nobel laureates in physics have all confirmed that Einstein and Special Relativity are right?"

Tuan was stuck. He didn't know how to answer.

"You ask me a question which I'm not qualified to answer. First, I'm not them. So, I can only guess. Second, despite its publicity, I don't think Special Relativity is the only interest for all physicists. Scientists have many other things they feel are more important, practical, or urgent to do. Third, since it's impossible to know everything these days, many specialists choose to focus on a very narrow field and rely on their colleagues' expertise in other fields. Einstein has never attempted to measure light speed. That is not his expertise. He relies on others. Others rely on him with the absolute light-speed postulate without questioning where it came from. We test time dilation relying on perfect atomic clocks made by other clock specialists. But atomic clocks, according to their creators, are not perfect and can become faulty outside their working constraints. Lost-in-translation information does happen. We accidentally mistrust, misinterpret, or misuse information delivered by other specialists. Last, anyone can make a mistake, even a genius. If Einstein could, anyone could."

"Let's talk about your qualifications. Apart from a high school certificate, do you have anything else?"

"If I agreed with Einstein, would you ask for my qualifications?"

"Just answer my questions. Do you have a PhD?"

"No."

"Any degree in physics or science?"

"No."

"Any TAFE qualifications?"

"No."

"Then how dare you talk about Special Relativity without any scientific qualifications against all those physics professors, Nobel laureates, and respected scientists, whom you call heroes and great minds? Don't you think, by attacking them, you've made a fool of yourself and a laughing stock of yourself?"

"First, I didn't know when to question time dilation became illegal. Second, you've made it sound like I was attacking all my heroes, the likes of Einstein, Hawking, Feynman, and all the other great scientists. It is a wrong accusation. If you replay the evidence all my friends gave in this trial, none of us made any personal attacks on any scientists. We attack no one. We respect all scientists. We admire them. We only question a potential error. Why does $c + v = c$? Why does $1 + 2 = 2$? And if Special Relativity cannot explain, we have no choice but to say it's nonsense. That's exactly what we've learned from our kindergarten teachers to say: $1 + 2 = 2$ is nonsense. Last, regarding your question about being a laughing stock, if I can make someone laugh, I think it's a good thing. Life is short. Laugh while you can."


"You seem to contradict yourself. Did you say before that you don't have enough qualifications and that you need to learn more?"

"Yes."

"Did you say that you always look up to many physicists, but their level is too high for you?"

"Yes."

"And that's why you don't always understand their works, right?"

"Yes, you're right. I don't always understand their works, but that doesn't mean what I understand is wrong."

"But you must admit one of the reasons for your failure to understand Special Relativity is because Special Relativity is too high for you. You said that before, didn't you?"

"No, I didn't say that. I said I don't always understand all physicists because their level is too high for me. But talking about Special Relativity as a theory dealing with space–time, length contraction, time dilation, absolute light speed, simultaneity, and so on, I don't think it's too high for me. They teach us in high school for a reason."

"What about math? Do you understand all the math behind Special Relativity?"

"No."

"Good grief. Now you admit that you don't understand all the math behind Special Relativity?"

"Correct. I don't fully understand its math."

"This is silly. This is absurd. How can you criticize a scientific theory like Special Relativity when you don't fully understand its math? Don't you think what you've just said is absolute nonsense?"

"I don't think math can answer everything in life. If it could, why should we learn other subjects in schools? Math is just another language used for communication, like Vietnamese, Russian, or any other language. It's compact, concise, and complete; therefore it's a very useful and powerful tool to explain your logic. But math does not necessarily mean reality, as Einstein said: *As far as the laws of mathematics refer to reality, they are not certain; and as far as they are certain, they do not refer to reality.* Our history has many examples of math misuse resulting in wrong predictions. Like the story of Anaxagoras, who, as told by one of my friends earlier, used trigonometry some centuries ago to wrongly calculate the Sun–Earth distance as 6500 km, but it turned out it's Earth's radius. True mathematicians know math has limits. Math is never wrong, but applying math to reality requires assumption. If our assumption is right, it works wonders. But if our assumption is wrong, it's human error, not math."

"What's wrong with Einstein's assumption?"

"The second postulate. Einstein assumed the constancy of light speed regardless of the observers' motion. There's no logic in it. It makes no sense. He just made it up. Since then, all math in Special Relativity has been manipulated to create a veneer of validation for that wrong assumption. Nothing else."

"How can you say that when you don't fully understand its math?"

"I've already answered. I don't know what else I can say."

"I tell you why you cannot understand Special Relativity. There's nothing wrong with Special Relativity, but your poor math is your failure to understand Special Relativity. What do you say about that?"

Tuan hesitated. He tried to figure out a way to explain. "Do you play tennis, Sir?"

"Don't ask me."

"Sorry. Let's just say halfway through a game in a tennis match, I'm about to serve and just forget the score. Then someone yells out 15–30. Even though I completely forget the score, I can still tell that is wrong, for I'm standing on the deuce side – either the score is 15–15, 30–30, 40–40, 40–15, 15–40, but it cannot be 15–30. So you see, sometimes in life, you can tell something is wrong even though you don't know everything."

"I'm talking about math; why are you talking about tennis?"

"Well, it's the same principle. Let me try again. Suppose we need to calculate the value of a math product consisting of 100 factors. Many factors therein contain very advanced math equations way beyond my level, and I can't solve them. Except for the last factor, which is quite simple, and I can easily work out its value as zero. Since anything times zero equals zero, I immediately know the whole product's answer as zero without the need to calculate the rest. It's the beauty of math. Right is right, wrong is wrong, even if it's junior math. It's problem-solving using math logic. Sometimes, you do not need to know everything, but by knowing just one key factor well enough, you can solve the whole problem. The same thing happens with Special Relativity. I don't fully understand all complex math behind it, but I know math just enough to conclude that Special Relativity is wrong."

"Which is?"

"The second postulate of constant light speed regardless of observers' motion. If light has a constant speed, that speed must change if someone moves closer to or away from that light. That is the definition of speed. Measurement is to assign one number to one thing. We cannot assign two different numbers to one thing; otherwise, two different numbers are the same. In science, one thing can only have one measurement. Denying it means faulty measurement. That's exactly what Einstein did. He tried to validate that faulty measurement into a postulate, and to make $c + v = c$ add up, Einstein had to alter the value of time – a measurement unit. By doing so, he had violated the most fundamental rule of science – never change measurement units or measurements make no sense. You cannot do math if you change the value of numbers. You cannot measure if you alter measurement units. You cannot make sense if you deny sense. Without absolute measurement systems, without absolute rulers, without absolute clocks, all Special Relativity calculations and measurements cannot be trusted. They're roots for confusion. They're causes for illusion. They're recipes for disasters."

"What helps? What can we do to convince you that you're wrong?"
"That should be easy. Just show me where I'm wrong."
"How?"
"Answer my list of 10 questions."
"Is that all?"
"Yes, with only 10 words."
"How?"
"Just answer yes or no."
"Have you asked anyone else these questions?"
"Yes. We've asked our teachers. They're not sure. We've asked all the physicists that we can contact. They ignore us. They avoid our questions. They can only answer their questions, but not ours. The deadlock remains. Later on, we found out about other issues. Some physicists admit they have doubts but cannot criticize Special Relativity for fear of losing their jobs. It's sad for science. But what can you do? Even in this trial, you hear how those mainstream physicists

responded to our barristers' questions. Do you think they've convinced you?"

"Stop asking me."

"Sorry. But the fact is, as you've heard, their answers are so inconsistent. The public had never heard of this inconsistency until today, when those experts faced scrutiny. I mean, the question is not why I cannot convince physicists, but why physicists cannot convince me. The question is not why I can't understand their advanced math, but why they cannot understand my simple language. The question is not why users cannot convince experts, but why experts cannot convince users. Why do I have to keep answering their questions, but they cannot answer mine? I mean, they're smarter than me. I'm only 17. I've just finished high school. I know what I know is not enough, and I need to learn more. If anyone can make a mistake, I'm probably the one who most likely makes more mistakes than anyone else. I mean, it should be very easy to convince me that I'm wrong. I know that. That's why I'm here – waiting for that correction. I'm waiting for that one hero who can come and open my eyes. My question is simple. If those experts know where I'm wrong, they can easily point it out. Why don't they?"

"You know $E = mc^2$?"

"Yes, the most beautiful equation in physics."

"Are you trying to dishonor that equation as well?"

"No. What does it have to do with what we're doing here?"

"If Einstein's absolute light speed is wrong, as you claim, what will happen to that equation?"

"I don't know. You ask me about things I don't know. I can only tell you what I know. And what I know is time dilation is wrong."

"But what about nuclear bombs? They're real, aren't they? If Einstein's Special Relativity is wrong, where do they come from?"

"No. Nuclear bombs are the product of the theory of fission, in which Einstein's equation is not involved. It's electrostatic repulsion between protons when—"

"Hang on, let me stop you there. You don't have to go over those technical details given by your friends that have already been disproven here. Let me put you in perspective. Think about this. There are hundreds of thousands of physicists in the world. Are you saying they're all wrong as well? And what about millions of science students in the world who study Einstein, admire him, and accept Special Relativity? Is the whole world wrong, and only you are right? C'mon, give me a break, will you? Who the hell do you think you are? Why are you so negative in opposing the progress science has made in life? Don't you appreciate everything you enjoy in life is possible thanks to science? Why are you so against the whole world?"

"No, I'm not against the world, science, or anyone. I only try to find truth for myself the same way everyone else does. In fact, we're all here for a better world, a better science, and a better future. We all have the right to see the true world. I believe we've been fooled by Special Relativity with its confusing language about time dilation, length contraction, and absolute light speed for too long.

"I think it's wrong. It's nonsense. It needs to be stopped. If a high school student like me can understand this, and understand enough to say this, anyone can. Regardless, scientists, students, lawyers, doctors, nurses, accountants, journalists, actors, anyone with standard education and some interest in the subject can see that time dilation is wrong. Time dilation is faulty time by faulty clocks. Special Relativity is a faulty theory based on faulty description and faulty measurement of the world. It's a theory of illusion. Even Einstein, my genius and hero, if he was here, I believe I could convince him. I would put to him all my questions, and by answering them, he would realize he was wrong. It's just a tiny mistake, as previously said by our barrister; no way would that affect his other greater achievements.

"I know that sounds impossible, and it's only impossible because we think it is. We think it's impossible that Einstein is wrong. We think it's impossible to defend common sense against math. We think that since there is no way we can understand superior math, we presume it's right. And we forget Einstein's warning that math is not reality. We forget we're masters of our destiny, and all theories, knowledge, and math are only vehicles we can choose to take us somewhere.

"But we can change all that. This is a new era, when knowledge is not only top-down or bottom-up but open dialogue with everyone. No one knows everything. And no one knows nothing. Users have as much of a say as scientists in what matters. Clients have much of a say as experts in what they pay for use. Experts cannot get better without clients' input. And clients cannot get better without experts' output. We need to learn from everyone.

"After all, Einstein is not God. He's only a genius. Adore him for his greatness. But question him for problems. He's not perfect and he needs our questions for further perfection. Our silence means approval. Our blind worship stops him from becoming even greater and makes us stupid followers. No one wants that. It doesn't benefit the world. The world needs all of us – to be free of illusion, misconception, and misinformation.

"Change must start with science. Science must lead. Science must not falsify. Physics needs *perestroika*. Scientists need to escape the illusion of absolute light speed as maximum calculation. It's the wall blocking our view and distorting our vision. Let's tear it down. Let's go beyond 'c' and discover new physics. We need to see the true world first, before we can make it a better one."

Mr. McNamara couldn't hide his amusement. He chuckled at the rhetoric full of fancy, aspiring, but empty words, then quickly changed his tone and direction for a final assault. "Have you ever been wrong in high school, young man?"

"Yes."

"At the time when you're wrong, would you know?"

"No."

"Then how could you find out eventually?"

"Someone would point out and let me see."

"What happens after you've realized you were wrong?"

"I said sorry. I learn from my mistakes. I try to be better next time."

"Would you sometimes feel embarrassed or sad about mistakes you've made?"

"Yes."

"And wish you had not made them?"

"Yes."

"I mean, it's normal, isn't it? You've just finished high school. You're only 17. You said what you know is not enough, that you need to learn more, and that you're probably the one who most likely makes more mistakes than anyone else. Do you agree that's exactly what you've just said?"

"Yes, absolutely."

"Yes, absolutely. And that could absolutely happen here again. I put it to you that there's nothing wrong with Special Relativity. It's just you who cannot understand it because of your low level. But one day, hopefully, if you become a bit smarter, you will realize your mistakes. You would feel embarrassed for doing all this, wasting your time, your friends' time, and everyone's time. Instead, you could've used that time to do things wiser, like studying more science, finding a better job than cleaning toilets, or saving more money to help your friends in Vietnam who cannot shower because it's too cold. There's a limit to what scientists can do. There are millions of scientists out there who are so busy doing more important work for the community than to waste time convincing some poor kids like you. There are millions of tertiary students in the world who are learning more important things than wasting their time explaining a theory proven for over a century that only crackpots now would try to challenge it. You don't deserve their time. Your group of 21 students is only a tiny minority against the whole world of educated and smart people. My only advice for you is to go back to school and learn more today, so that tomorrow you can have a better vision and a better future. Until then, like someone with a disability, you just cannot see now. Am I right?"

"Yes."

"I've got no further questions," Mr. McNamara said.

It was heavy and nasty. Everyone in the courtroom anxiously looked at the boy. Suddenly, he looked vulnerable. Is that how relativity was taught and learned in colleges through that verbal thuggery? Tuan felt tired. His barrister could've interfered by objecting to those malicious questions but chose not to. She sensed a fiery resilience from the boy despite his fragile look and decided to leave him alone. People get beaten to death in war, though all they ask is a reason to believe. After they fall, their fight goes on. She'd seen that. He'd seen that. This is nothing.

Finally, it was the chief judge who interrupted.

"You might be wrong, but you're very smart. You could fail, but you're very brave. But what did you mean by saying 'yes' to the last question?"

Tuan tried to hold up his head using his hand, as often he did so during the trial, as if it were about to fall off. No one knows he was just a ghost hanging on life for some unfinished business. Then he said, his last words about a long-lost dream, only few would understand. (See Figure 17.1.)

"I think he's right. Everyone has a disability. Visible or hidden, it's something that stops us from achieving more. You can deny it and let it block you. Or see it, face it, climb over it, and leap higher.

"After all, I've said nothing new. I've made no discovery. This is what I remember, what I think, from everything I've learned. If my knowledge is not good enough, I'm sorry, I can only try. Some oceans are too far for me to reach. Whatever, I love my teachers. They all tell me to never fake. Always be true, even if you fail.

"Right or wrong, I still think Einstein is bigger than Relativity. Science is bigger than Einstein. And humanity is bigger than science. What matters is no longer Relativity but the purpose of knowledge, and what better way we can do for each other as humans. Anyone can claim he knows, but what his knowledge can do is what counts. Would it improve life for many or only boost his ego by treating everyone else as stupid? Who knows, Relativity could be just a joke from Einstein's imagination to test how smart, how foolish, and how courageous we all can be?

"I understand that when we grow up, life changes. Things we learn in schools today may be different from what life teaches us tomorrow. People need jobs, have bills to pay, want money to spend, and demand titles, status, and crowns to fit in. Life's equation becomes messier with other variables like fear, fake, and fame. Some fear ridicule. Some fake to impress. Some need fame for power. That could change people, cloud their thinking, and alter what they say. I blame no scientists. But as a poor student who just wants to learn and search for truth, I just hope one day a true scientist somewhere out there can go beyond those limits and reach me – and liberate me. I'm from the other side. But I'm not a bad guy. I'm not an enemy.

"We were all seventeen once. There's one schoolboy in every one of us. Set him free."

# EPILOGUE
## OTHER SIDE OF THE SUN

*Education is not the learning of facts,*
*but the training of the mind to think.*
*ALBERT EINSTEIN*

It looks like another world. I walked into one of the world's most prestigious universities and suddenly felt belittled. In front of me was a professor, the Head of the School of Physics. He was busy reading something and must have forgotten me. After another few minutes, I decided to remind him by softly clearing my throat. He was still reading, though. Finally, he dropped his glasses, slowly gazed through the window to the sun outside, and asked, "What's on the other side of the sun?"

I frowned, unsure what he meant or whom he asked. But before I could say anything, he'd already said something else.

"I need a fake book – one that sounds very much anti-Einstein and anti-Relativity. Special Relativity is a monumental theory in modern physics, already approved by all professional physicists and studied by millions of university students around the world. We have an enormous amount of evidence confirming nearly every aspect of Einstein's Relativity, data from LHAASO, atomic falls, Lorentz symmetry, the LHC test, you name it. Unfortunately, some people are yet to understand. The theory requires some advanced math and unconventional reasoning that not everyone is capable of. That happens to some of my struggling students, too. Including some people who pretend to understand the theory, but in fact they don't. You ask them a few silly questions, and they get confused. A few more, and they disappear. They can repeat very well, but they can't reason. They can explain very well, but they can't answer. They can talk very well, but they have no idea what they're talking about. As an educator, I believe there must be a better way to teach knowledge, test knowledge, and discover knowledge. So one day, I was thinking. Instead of giving them textbooks to copy, or feeding them with words to repeat, why not give them a faulty fiction from the other side and see how they would respond? Maybe by facing wrong questions, doubts, and illusions, they might find their own reasoning and discover the truth for themselves. Can you write me a novel?"

THE END

# APPENDIX:
# THE LIST

1. Relativity is essentially based on the second postulate of absolute light speed regardless of the source's and observers' motion. If this postulate is found to be wrong, invalid, or senseless, the whole theory, its velocity addition formula, and its two equations of time dilation and length contraction will fall. True or false?

2. According to Relativity's second postulate, if two light beams travel in opposite directions, the relative speed of one beam to the other is always "c," not "2c;" and if two beams travel alongside in the same direction, the relative speed of one beam to the other is always "c," not zero. Is that what Relativity says based on its velocity addition formula?

3. In the second case of two light beams traveling alongside in the same direction, if each beam carries a driver, can those drivers shake hands during the trip because they're side by side?

4. If two friends can hold hands during the whole trip, their distance from each other must remain the same, and their relative speed to each other must be zero, according to the definition of speed. True or false?

5. Does that mean Relativity's second postulate violates the definition of speed and falsifies measurement?

6. Distance requires two points. We cannot say A is 10 km, but A–B is 10 km, A–C is 20 km, A–D is 45 km, and so on. In math, each number is unique, different, and never repeats. Likewise, every point in space has a unique position. No point can have the same distance to everything else in space. True or false?

7. Speed, as distance over time, requires two points. We cannot say something has a speed without reference to something else. When we say a car's speed is 100 km/h, we mean its speed relative to a certain point on the ground. As Earth rotates around its axis, Earth also revolves around the Sun, the Sun also revolves around its galaxy's center, and millions of things are moving in different directions at different speeds in space. A car's speed can be all different relative to everything else in space. Nothing can move the same distance to everything else in space at any time; therefore, nothing can have the same speed relative to everything. True or false?

8. To validate the second postulate, we can change the value of space and time to make "c" here look different from "c" over there, one mile here different from one mile over there, one second here different from one second over there, but would that violate the rule of measurement, which states the same things must have the same measurement and different things have different measurements?

9. Science is to make a difference. Measurement is required to make sure the difference is not fake, but real. Measurement works by comparing everything with the same thing called measurement units. But if scientists invent time dilation and length contraction to alter the length of one second and one mile without realizing they're altering measurement units, use slow clocks to prove time dilation without realizing they're using faulty clocks to produce faulty measurements, and confuse themselves by using words with different meanings, and numbers with different values,

would you trust them to make a real difference, and not fake?

.

*Figure 18.1 A point alone cannot have distance or speed*

10. Relativity starts with the absolute light speed postulate but fails the definition of speed. Relativity ends with time dilation but fails the purpose of time. Is the end of time dilation the end of Relativity? (See Figure 18.1.)